T0213274

JOINT COGNITIVE SYSTEMS

SYSTEMS

Foundations of
Cognitive Systems Engineering

JOINT COGNITIVE SYSTEMS

SYSTEMS

Foundations of
Cognitive Systems Engineering

Erik Hollnagel
David D. Woods

CRC Press
Taylor & Francis Group
Boca Raton London New York

CRC Press is an imprint of the
Taylor & Francis Group, an **informa** business
A TAYLOR & FRANCIS BOOK

CRC Press
Taylor & Francis Group
6000 Broken Sound Parkway NW, Suite 300
Boca Raton, FL 33487-2742

First issued in paperback 2019

ISBN-13: 978-0-8493-2821-8 (hbk)
ISBN-13: 978-0-367-86420-0 (pbk)

Library of Congress Cataloging-in-Publication Data

Hollnagel, Erik, 1941-
   Joint cognitive systems : foundations of cognitive systems engineering / Erik Hollnagel and David D. Woods.
         p. cm.
   Includes bibliographical references and index.
   ISBN 0-8493-2821-7 (alk. paper)
   1. Human-machine systems. I. Woods, David D. II. Title.

TA167.H67 2005
620.8'2--dc22   2004064949

Visit the Taylor & Francis Web site at
http://www.taylorandfrancis.com

and the CRC Press Web site at
http://www.crcpress.com

Contents

x

Preface

"At any one time a science is simply what its researches yield, and the researches are nothing more than those problems for which effective methods have been found and for which the times are ready."
Edwin G. Boring (American psychologist, 1886-1968)

Deciding what to write in a preface is not as easy as it seems. If there is anything of substance to say about the subject matter, it should clearly be in the main text of the book rather than in the preface. And if there is nothing of substance to say, then why say it? Prefaces can therefore easily become the place where the authors write their more personal views on matters large and small, express their gratitude to various people, and lament about the state of the world in general. Prefaces can also become the authors' encomium for their own work. Publishers welcome that, since it can be put on the back cover or on the website, but common decency usually intervenes in time.

After this beginning there are pretty few options left for us, except to briefly reminisce on how this book came about. The collaboration between the two authors began around 1979 and soon led to the ideas that became formulated as cognitive systems engineering (CSE), first in an internal report in 1982, and later as a paper in the *International Journal of Man-Machine Studies*. About ten years later, in the beginning of the 1990s, we started talking about writing a book on CSE. One motivation was that the idea seemed to have caught on; another was that we by that time individually and together had produced a number of writings that both had developed the original ideas further and illustrated how they could be applied in practice.

The route from word to deed is, however, usually longer than initially hoped for. While the intention to write a book together was never abandoned, progress was painfully slow, partly because we both had entangled ourselves in too much interesting work, and partly because the much coveted opportunity to sit down together for a couple of weeks on a desert island never quite materialised.

In the end the book became a reality because we adopted a pragmatic solution – in good according with the ethos of CSE. Instead of writing one book together, the project was split into two parts, leading to two books that complement each other. While we appear as joint authors of either book, the order differs to reflect the fact that the first author in each case is the main responsible for the writing. This also helps to solve the practical problem that a single book would have been rather large. Furthermore, we had over the years developed different styles of working and writing, one being more contemplative, the other being more practical.

The end result is therefore not one but two books of which this is the first. Both books are entitled *Joint Cognitive Systems* with the subtitles indicating the specific focus – and therefore also what distinguishes them from each other. The subtitle of the present book is 'Foundations of Cognitive Systems Engineering', while the subtitle of the second book is 'Patterns in Cognitive Systems Engineering'. The intention is that either book can be read independently, but that they also will complement each other by emphasising the theoretical and the practical aspects respectively. It is no requirement that they are read in a specific order. Yet if people after reading either one would find it necessary to read also its counterpart, we would feel we had achieved our purpose.

Erik Hollnagel   David D. Woods

Chapter 1

The Driving Forces

The focus of CSE is how humans can cope with and master the complexity of processes and technological environments, initially in work contexts but increasingly also in every other aspect of daily life. The complexity of the current technological environment is not only something that must be mastered but paradoxically also provides the basis for the ability to do so. This entangling of goals and means is mirrored in the very concepts and theories by which we try to understand the situation of humans at work. To set the context, this chapter gives an overview of the scientific developments of the 20th century that have shaped our thinking about humans and machines.

INTRODUCTION

This book could reasonably start by asking how Cognitive Systems Engineering – in the following abbreviated to CSE – came about. Yet more relevant than accounting for the *how* is accounting for the *why*, by describing the forces that led to the formulation of the basic ideas of CSE. Such a description is important both to understand what CSE is all about, and as a justification for the CSE as it is today. Although CSE was formulated more than twenty years ago (Hollnagel & Woods, 1983), the basic message is certainly not outdated and its full potential has not yet been realised.

A description of the driving forces has the distinct advantage of being based on hindsight, which makes it possible to emphasise some lines of development that are useful to understand the current situation, without blatantly rewriting history as such. The three main driving forces are listed below. As the discussion will show, the situation at the beginning of the 21st century is in many ways similar to the situation at the end of the 1970s.

- The first driving force was the growing complexity of socio-technical systems, which was due to the unprecedented and almost unrestrained growth in the power of technology, epitomised by what we now call computerisation or applied information technology. This development

1

started slowly in the 1930s and 1940s, but soon after gained speed and momentum so that by the end of the 1970s computers were poised to become the dominating medium for work, communication, and interaction – revolutionising work and creating new fields of activity.

- The second driving force was the problems and failures created by a clumsy use of the emerging technologies. The rapid changes worsened the conditions for already beleaguered practitioners who often had insufficient time to adjust to the imposed complexity. One consequence was a succession of real world failures of complex systems that made human factors, human actions, and in particular the apocryphal 'human error' more conspicuous.
- The third driving force was limitations of linear models and the information processing paradigm. Although the popularity of human-computer interaction was yet at an initial stage, the view of humans as information processing systems had been keenly adopted by the engineering and computer science communities, leading to a fragmentary view of human-machine interaction.

Computerisation itself was the outcome of a number of theoretical and technological developments that went further back in time. Some of these were helpful and provided the concepts, models and methods that made it possible to address the practical issues of the time – around the middle of the 20th century – while others were decidedly unhelpful, in the sense that they accidentally created many of the problems that practitioners had to struggle with. It is usually the case that the positive aspects of an innovation – be they technical or conceptual – attract attention and therefore often quickly are used in applications. In the initial enthusiasm the negative aspects are easily overlooked and therefore only become clear later – sometimes even much later. This may happen in a concrete or material sense such as with nuclear power or the general pollution caused by industrial production. It may also happen in an incorporeal sense such as when a certain way of thinking – a certain paradigm in the Kuhnian meaning of the term (Kuhn, 1970) – turns out to be a stumbling block for further development. As we shall argue throughout this book, the information processing paradigm represents such as case. The positive sides were immediately and eagerly seized upon, but the negative sides only became clear almost half a century later.

On Terminology

Before proceeding further, a few words on terminology are required. Due to the background and tradition of CSE, the focus is on human-machine systems rather than human-computer systems, where the term *machine* is interpreted broadly as representing any artefact designed for a specific use. For the same reasons, the human in the system is normally referred to as an *operator* or a *practitioner*, rather than a user. Finally, a system is used broadly to mean the deliberate arrangement of parts (e.g., components, people, functions, subsystems) that are instrumental in achieving specified and required goals (e.g., Beer, 1964).

COMPUTERISATION AND GROWING COMPLEXITY

When we refer to a technological system, we invariably think of it in the context of its use. People are therefore always present in some way, not just in the sense of individuals – such as in the paradigmatic notion of the human-machine system – but also in the sense of groups and organisations, i.e., social systems. Technological systems are of interest because of how they are used, rather than as pieces of equipment that exists physically – made up of mechanical, electronic, hydraulic, and software components. Regardless of whether the application is autonomous – as in the case of a space probe or a deep-sea robot – or interactive (with all kinds of shadings in between), a technological system is always embedded in a socio-technical context. Every system has been designed, constructed, tested, and put into use by people. Every system requires maintenance and repair, although for some it may be practically impossible to do so. Every system produces something, or represents something, with an intended use, hence with an intended user. In system design, people apply all their powers of creativity and imagination to prepare for the eventual application and to guard against possible failures.

Although all systems thus in a fundamental sense are socio-technical systems, it is useful to distinguish between *technological system*, where technology plays a central role in determining what happens, and *organisations*, where humans mainly determine what happens. In CSE, organisations are themselves considered as artefacts, as something devised for a specific purpose, although they are of a social rather than physical nature.

Self-Reinforcing Complexity Cycle

The intertwining of technology and complexity can be illustrated as in Figure 1.1. An arbitrary starting point for the cycle is the technology potential, which can be used to modify the way things are done as well as to introduce new functions altogether. Some familiar examples are the use of numerically controlled machines, industrial robots, computer-assisted design, flexible manufacturing, office automation, electronic exchange of data and funds, decision support systems, and the Internet. The growing technology potential is invariably seized upon and exploited to meet performance goals or efficiency pressures. This is referred to as the Law of Stretched Systems, originally suggested by Lawrence Hirschhorn:

Under resource pressure, the benefits of change are taken in increased productivity, pushing the system back to the edge of the performance envelope. (Woods & Cook, 2002, p. 141)

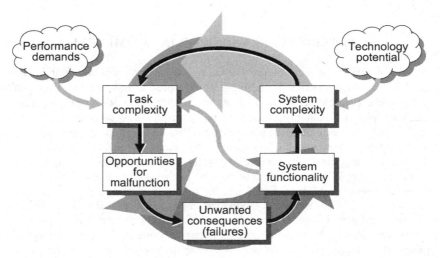

Figure 1.1: The self-reinforcing complexity cycle.

Some of the explicit motivations for putting technology to use are reduced production costs, improved product quality, greater flexibility of services, and faster production and maintenance. It need hardly be pointed out that these benefits are far from certain and that a benefit in one area often is matched by new and unexpected problems in another. Furthermore, once the technology potential is put to use this generally leads to increased system complexity. Although this rarely is the intended outcome, it is a seemingly inescapable side effect of improved efficiency or versatility. The increased system complexity invariably leads to increased task complexity, among other things (e.g., Perrow, 1984). This may seem to be something of a paradox, especially when it is considered that technological innovations often – purportedly – are introduced to make it easier for the users. The paradox is, however, not very deep and is just another version of the well-known irony of automation (Bainbridge, 1983). The growing task complexity generally comes about because adding functionality to a system means that there is an overall increase in complexity, even if there are isolated improvements. Another version of the paradox is the substitution myth, which will be discussed in Chapter 5.

As it can be seen from Figure 1.1, growing task complexity may also be a result of increased performance demands. Since our technological systems – including, one might add, our social institutions – often seem to be working at the edge of their capacity, technology potential provides a way to increase capacity. This may optimistically be done in the hope of creating some slack or a capacity buffer in the system, thereby making it less vulnerable to internal or external disturbances. Unfortunately, the result invariably seems to be that system performance increases to take up the new capacity, hence bringing the system to the limit once more. Consider, for instance, the technological developments in cars and in traffic systems. If highway driving today took place at the speed of the 1940s, it would be very safe and very comfortable. Unfortunately, the highways are filled with more drivers that want to go faster, which means that driving has become more complex and more risky for the individual. (More dire examples can be found in power production industries, aviation, surgery, etc.)

Complexity and Unpredictability

Continuing through Figure 1.1, increasing task complexity together with increasing system complexity means more opportunities for malfunctions. By this we do not mean just more opportunities for humans to make mistakes, but rather more cases where actions have unexpected and adverse consequences or where the whole system malfunctions. Consider, for instance, the Space Shuttle. In addition to the Challenger and Columbia accidents, it seems to be the rule rather than the exception that a launch is delayed, in most cases due to smaller failures or glitches. Clearly, the Space Shuttle as a system has become so complex that it is close to being unmanageable (Seife, 1999, p. 5).

The larger number of opportunities for malfunction will inevitably lead to more actual malfunctions, more failures, and more accidents, which close the circle, as shown in Figure 1.1. The increase may either be in the frequency of already known malfunctions or in the appearance of new types of failure. The increasing number of failures in complex systems is a major concern for the industrialised societies and has itself been the motivation for numerous developments in the methods used to analyse, prevent, and predict such accidents (Hollnagel, 2004). At present, we will just note that a common response to incidents and accidents is to change the functionality of the system, typically by introducing additional barriers and defences. Although this in principle can be done without making the system more complex – in fact, it can sometimes be done by making the system simpler – the general trend is to add technology to make systems safer. Very often technological developments, such as increased automation, are promoted as the universal solution to accident prevention (the aviation industry being a good example of that). Another common solution is to introduce new barrier functions and defences to avoid future accidents, thereby making the systems more complex.

The net result of these developments is a positive feedback loop, which means that deviations will tend to grow larger and larger – resulting in more serious events and accidents (Maruyama, 1963). Although this interpretation may be overly pessimistic, the fact remains that we seem to be caught in a vicious circle that drives the development towards increasingly complex systems. One of the more eloquent warnings against this development was given in Charles Perrow's aptly named book *Normal Accidents* (Perrow, 1984), in which he argued that systems had become so complex, that accidents were the norm rather than the exception. It is in this sense that the growing technological complexity is a challenge as well as a motivation for CSE.

Figure 1.1 is obviously a simplification, which leaves out many nuances and opportunities for self-correction in the loop. In reality the situation is not always as bad as Figure 1.1 implies, since most complex systems function reliably for extended periods of time. Figure 1.1 is nevertheless useful to illustrate several important points.

- Systems and issues are coupled rather than independent. If we disregard these couplings in the design and analysis of these systems, we do it at our own peril.
- Events and relations must be understood in the context where they occur. It is always necessary to consider both dependencies to other parts of the system and to events that went before. This is particularly the case for human activities, which cannot be understood only as reactions to events.
- Control is fundamental in the definition of a cognitive system. Since all systems exist in environments that to some extent are unpredictable, there will sooner or later be a situation that was not considered when the system was designed. It can be difficult enough to keep control of the system when it is subject only to the 'normal' variability of the environment, but it becomes a serious challenge when unexpected situations occur. In order for the system to continue to function and maintain its integrity, control is necessary whether it is accomplished by the system itself or by an external agent or entity.

The positive feedback loop described above is a useful basis for understanding changes in human interaction with technology. Chapter 2 will provide a more thorough treatment of this issue and describe how technological developments have led to changes in the nature of work. For the moment we shall simply name three significant consequences of the growing complexity.

- The striving for higher efficiency inevitably brings the system closer to the limits for safe performance. Concerns for safety loom large and neither public opinion nor business common sense will accept efficiency gains if they lead to significantly higher risks – although there sometimes may be different opinions about when a risk becomes significantly higher. Larger risks are countered by applying various kinds of automated safety and warning systems, although these in turn may increase the complexity of the system hence lead to even greater overall risks. It may well be that the number of accidents remains constant, but the consequences of an accident, when it occurs, will be more severe.

- A second important issue is the increased dependence on proper system performance. If one system fails, it may have consequences that go far beyond the narrow work environment. The increasing coupling and dependency among systems means that the concerns for human interaction with technology must be extended from operation to cover also design, implementation, management, and maintenance. This defines new demands to the models and methods for describing this interaction, hence to the science behind it.

- A third issue is that the amount of data has increased significantly. The sheer number of systems has increased and so has the amount of data that can be got from each system, due to improved measurement technology, improved transmission capacity, etc. Computers have not only helped us to produce more data but have also given us more flexibility in storing, transforming, transmitting and presenting the data. This has by itself created a need for better ways of describing humans, machines, and how they can work together. Yet although measurements and data are needed to control, understand, and predict system behaviour, data in itself is not sufficient. The belief that more data or information automatically leads to better decisions is probably one of the most unfortunate mistakes of the information society.

CONSPICUOUSNESS OF THE HUMAN FACTOR

Over the last 50 years or so the industrialised societies have experienced serious accidents with unfortunate regularity, leading to a growing realisation of the importance of the human factor (Reason, 1990). This is most easily seen in how accidents are explained, i.e., in what the dominant perceived causes appear to be.

It is commonly accepted that the contribution of human factors to accidents is between 70% – 90% across a variety of domains. As argued elsewhere (Hollnagel, 1998a), this represents the proportion of cases where the *attributed* cause in one way or another is human performance failure. The attributed cause may, however, be different from the actual cause. The estimates have furthermore changed significantly over the last 40 years or so, as illustrated by Figure 1.2. One trend has been a decrease in the number of accidents attributed to technological failures, partly due to a real increased reliability of technological systems. A second trend has been in increase in the number of accidents attributed to human performance failures, specifically to the chimerical 'human error'. Although this increase to some extent may be an artefact of the accident models that are being used, it is still too large to be ignored and probably represents a real change in the nature of work. During the 1990s a third trend has been a growing number of cases attributed to organisational factors. This trend represents a recognition of the distinction between failures at the sharp end and at the blunt end (Reason, 1990; Woods et al., 1994). While failures at the sharp end tend to be attributed to individuals, failures at the blunt end tend to be attributed to the organisation as a separate entity. There has, for instance, been much concern over issues such as safety culture and organisational pathogens, and a number of significant conceptual and methodological developments have been made (e.g. Westrum, 1993; Reason, 1997; Rochlin, 1986; Weick, Sutcliffe & Obstfeld, 1999).

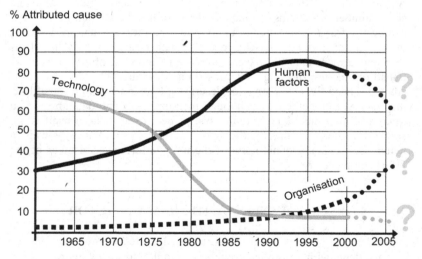

Figure 1.2: Changes to attributed causes of accidents.

The search for human failures, or human performance failure, is a pervasive characteristic of the common reaction to accidents (Hollnagel, 2004). As noted already by Perrow (1984), the search for human failure is the normal reaction to accidents:

> Formal accident investigations usually start with an assumption that the operator must have failed, and if this attribution can be made, that is the end of serious inquiry. (Perrow, 1984, p. 146)

Since no system has ever built itself, since few systems operate by themselves, and since no systems maintain themselves, the search for a human in the path of failure is bound to succeed. If not found directly at the sharp end – as a 'human error' or unsafe act – it can usually be found a few steps back. The assumption that humans have failed therefore always vindicates itself. The search for a human-related cause is reinforced both by past successes and by the fact that most accident analysis methods put human failure at the very top of the hierarchy, i.e., as among the first causes to be investigated.

THE CONSTRAINING PARADIGM

One prerequisite for being able to address the problems of humans and technological artefacts working together is the possession of a proper language. The development of a powerful descriptive language is a fundamental concern for any field of science. Basically, the language entails a set of categories as well as the rules for using them (the vocabulary, the syntax, and the semantics). Languages may be highly formalised, as in mathematics, or more pragmatic, as in sociology. In the case of humans and machines, i.e., joint cognitive systems, we must have categories that enable us to describe the functional characteristics of such systems and rules that tell us how to use those categories correctly. The strength of a scientific language comes from the concepts that are used and in precision of the interpretation (or in other words, in the lack of ambiguity). The third driving force of CSE was the need of a language to describe human-technology coagency that met three important criteria:

- It must describe important or salient functional characteristics of joint human-machine systems, over and above what can be provided by the technical descriptions.
- It must be applicable for specific purposes such as analysis, design, and evaluation – but not necessarily explanation and theory building.
- It must allow a practically unambiguous use within a group of people, i.e., the scientists and practitioners who work broadly with joint human-machine systems,

In trying to describe the functioning and structure of something we cannot see, the mind, we obviously use the functioning or structure of something we can see, i.e., the physical world and specifically machines. The language of mechanical artefacts had slowly evolved in fields such as physics, engineering, and mechanics and had provided the basis for practically every description of human faculties throughout the ages. The widespread use of models borrowed from other sciences is, of course, not peculiar to psychology but rather a trait common to all developing sciences. Whenever a person seeks to understand something new, help is always taken in that which is already known in the form of metaphors or analogies (cf., Mihram, 1972).

Input-Output Models

The most important, and most pervasive, paradigm used to study and explain human behaviour is the S-O-R framework, which aims to describe how an organism responds to a stimulus. (The three letters stand for Stimulus, Organism, and Response.) The human condition is one of almost constant exposure to a bewildering pattern of stimuli, to which we respond in various ways. This may happen on the level of reflexes, such as the Patella reflex or the response of the parasympathetic nervous system to a sudden threat. It may happen in more sophisticated ways as when we respond to a telephone call or hear our name in a conversation (Cherry, 1953; Moray, 1959; Norman, 1976). And it happens as we try to keep a continued awareness and stay ahead of events, in order to remain in control of them.

Although the S-O-R paradigm is strongly associated with behaviourism, it still provides the basis for most description of human behaviour. In the case of minimal assumptions about what happens in the organism, the S-O-R paradigm is practically indistinguishable from the engineering concept of a black box (e.g. Arbib, 1964), whose functioning is known only from observing the relations between inputs and outputs. The human mind in one sense really is a black box, since we cannot observe what goes on in the minds of other people, but only how they respond or react to what happens. Yet in another sense the human mind is open to inspection, namely if we consider our own minds where each human being has a unique and privileged access (Morick, 1971).

That the S-O-R paradigm lives on in the view of the human as an information processing system (IPS) is seen from the tenets of computational psychology. According to this view, mental processes are considered as rigorously specifiable procedures and mental states as defined by their causal relations with sensory input, motor behaviour, and other mental states (e.g. Haugeland, 1985) – in other words as a Finite State Automaton. This corresponds to the *strong* view that the human *is* an IPS or a physical symbol system, which in turn 'has the necessary and sufficient means for general intelligent action' (Newell, 1980; Newell & Simon, 1972). The phrase 'necessary and sufficient' means that the strong view is considered adequate to explain general intelligent action and also implies that it is the only approach that has the necessary means to do so. In hindsight it is probably fair to say that the strong view was too strong.

The strong view has on several occasions been met by arguments that a human is *more than* an IPS and that there is a need of, for instance, intentionality (Searle, 1980) or 'thoughts and behaviour' (Weizenbaum,

1976). Interestingly enough, it is rarely doubted whether a human is *at least* an IPS or whether, in the words of cognitive science, cognition is computational. Yet for the purpose of describing and understanding humans working with technology, there is no need to make assumptions of what the inner mechanisms of cognition might be. It is far more important to describe what a cognitive system *does*, specifically how performance is controlled.

Although the S-O-R framework is no longer considered appropriate as a paradigm in psychology, it can still be found in the many models of human information processing and decision making that abound. It seems as if the study of human behaviour has had great difficulty in tearing itself away from the key notion that behaviour can be studied as the relation between stimulus (input) and response (output), even though the fundamental flaw of this view was pointed out more than one hundred years ago:

> The reflex arc idea, as commonly employed, is defective in that it assumes sensory stimulus and motor response as distinct psychical existences, while in reality they are always inside a co-ordination and have their significance purely from the part played in maintaining or reconstituting the co-ordination. (Dewey, 1896, p. 99)

Translated into the current terminology, Dewey made the point that we cannot understand human behaviour without taking into consideration the context or situation in which it takes place. Specifically, he made the point that it is wrong to treat the stimulus (input) and the response (output) as separate entities with an independent existence. They are both abstractions, which achieve their reality from the underlying paradigm and therefore artefacts of the paradigm. Consequently, if the principle of the S-O-R paradigm is abandoned, the need to focus on input and output becomes less important.

The Shannon-Weaver Communication Model

We are by now so used to the input-output model that we may no longer be aware of its peculiarities, its strengths, and its weaknesses. Seen from the perspective of the behavioural sciences, the ubiquitous graphical form of the input-output model can be traced to the seminal book on information theory by Shannon & Weaver (1969; org. 1949). The beginning of this book introduced a diagrammatic representation that, perhaps fortuitously, provided the sought for 'image' of the S-O-R model. Because of that we may refer to the Shannon-Weaver model as the 'mother of all models'.

As shown by Figure 1.3, the model illustrates how a sender, called an information source, generates a message. The message is changed into a signal by the transmitter and sent through an information channel to the receiver. In the receiver, the signal is again changed into a message, which finally reaches the destination. The communication model was originally developed to describe how something like a telephone system worked. If, however, we throw the telephone away the result is the prototypical situation of person A speaking to person B. The message is what person A, the sender, wants to say. The signal is the sound waves generated by the vocal system of the sender (the transmitter) and carried through the air (the channel) to the ear of person B (the receiver). The ears and the brain of the receiver transform the sounds into a meaning, which then (hopefully) has the desired effect on the receiver's behaviour. The description can obviously be applied in the reverse order, when the receiver becomes the sender and vice versa. This leads to the paradigmatic case of two-way communication.

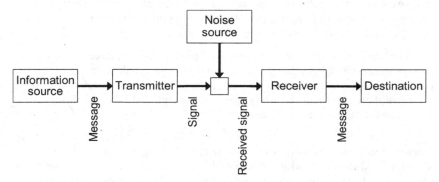

Figure 1.3: The Shannon-Weaver model of communication.

Prototypical Information Processing

The basic Shannon-Weaver model is doubled in Figure 1.4 to represent two-way communication. Basically the communication from sender to receiver is repeated, but now with person B, the original receiver, as the sender and person A, the original sender, as a receiver. This, of course, corresponds to the case of a dialogue, where – speaking more generally – system A and system B continuously changes their role. In order to make the first change 'work', so to speak, a second change has been introduced. This depicts how the message that has been received by system B is interpreted or processed internally, thereby giving rise to a new message that in the case of a dialogue is sent to system A, the original sender. The same change is also made for system A, who was the sender but who now has become the receiver.

It is not difficult to see the correspondence between this extended model and the common information processing model. Whereas information theory was interested in what happened to the signal as it was transmitted from sender to receiver, the emerging cognitive psychology was interested in what happened between receiving a message and generating a response. In this case the Shannon-Weaver model could be used to describe the internal processes of the mind as a series of transformations of information. An example of that is George Sperling's classical studies of auditory memory (Sperling, 1963 & 1967), which described how the incoming sounds were sent through a number of systems and transformed on the way until it reached the level of consciousness.

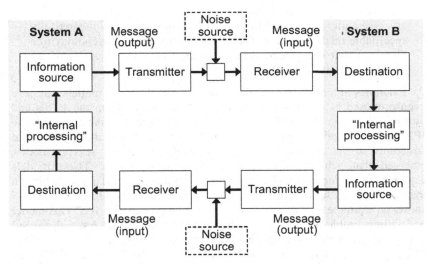

Figure 1.4: The extended Shannon-Weaver model.

In the present description there is no reason to go further into the details of how the model was developed. Several accounts have been provided of that (Attneave, 1959), although the basic model itself soon was taken for granted and therefore dropped out of sight. As we shall argue in the following, there were important consequences of adopting this paradigm, some of which are only now becoming clear. Whereas the Shannon-Weaver model is appropriate to describe the transmission of information between two systems, it is not necessarily equally appropriate to describe how two people communicate or two systems work together. The paradigm emphasises interaction, where CSE emphasises coagency. Although coagency requires interaction, it does not follow that it can be reduced to that.

FROM HUMAN-MACHINE INTERACTION TO JOINT SYSTEMS

One reason for the massive influence of the information processing paradigm was that it made it possible to describe what happened in the human mind between stimulus and response without being accused of resorting to mentalism. It may be difficult today to appreciate the attraction of this, but fifty years ago it was certainly a determining factor. This has been mentioned by George Mandler as one of the two characteristics that set cognitive psychology apart from previous schools; the other was the recognition that cognition and consciousness were different concepts with only a partial overlap.

> By considering the various subsystems within the mental organisation as essentially independent entities connected by theoretically specifiable relationships, it has opened up theoretical psychology to a pluralism that is in sharp contrast to the monolithic theories of the 1930s and 1940s. (Mandler, 1975, p. 12)

Examples of the IPS metaphor can be found in the many models that have been proposed as explanations of human performance from the 1970s and onwards. A large number of these were mostly *ad hoc* explanations, since the IPS metaphor made it tantalisingly simple to suggest new mechanisms and structures to fit current problems. Gradually, however, several models emerged that achieved the status of consensus models. These models captured something essential about human performance, and expressed it in a manner that was both comprehensible and tremendously useful. Figure 1.5 shows an example of a model based on the limited capacity IPS. This model has an impeccable pedigree, going back to the fundamental research carried out by people such as Broadbent (1958), Cherry (1957), and Moray (1970). The important feature of this model, as formulated by Wickens (1987), was to point out that the limited attention resources could be considered for different modalities and that there were specific high-compatibility links between stimulus formats and central processing operations.

Information processing models can have different levels of detail and sophistication in how they account for the O in the S-O-R paradigm. Common to them all is that they start by some external event or stimulus (process information or the evaluation of a routine action) and end by some kind of response. In an IPS model the internal mechanisms are typically described in far greater detail than in an S-O-R model – at some point focusing on the *O* almost to the exclusion of the *S* and the *R*. It is nevertheless easy to appreciate that they have two fundamental similarities: the sequential progression through the internal mechanism, and the dependency on a stimulus or event to start the processing. Neisser (1976) caricatured the classical information processing view by describing the stages as 'processing', 'more processing', and 'still more processing'. Despite the fact that this was done before most of the information processing models were formulated and gained general acceptance, Neisser's criticism had little practical effect.

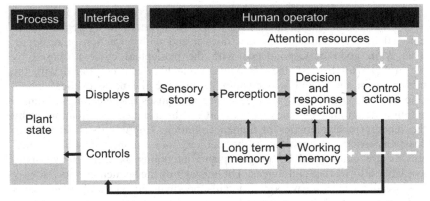

Figure 1.5: A limited capacity IPS model.

The Cognitive Viewpoint

In the 1980s, a growing number of people started to question the wisdom of considering the human as an IPS, in either the strong or the weak version. This led to the proposal of a model, which became known (in Europe, at least) as the cognitive viewpoint. This viewed cognition as active rather than re-active, as a set of self-sustained processes or functions that took place simultaneously, and changed the focus from the internal mechanisms of performance to overall performance as it could be observed. Behaviour was no longer seen as simply a function of input and system (mental) states, and the complexity of the inner 'mechanisms' was acknowledged to be too high to be captured adequately by a single theory.

The cognitive viewpoint introduced yet another change, namely the notion of the internal representation of the world in which the actions took place. This was formulated as follows:

> Any processing of information, whether perceptual or symbolic, is mediated by a system of categories or concepts which, for the information processing device, are a model of his (its) world. (De Mey, 1982, p. 84)

The internal representation, the system of categories, is characteristic for each system rather than generic and common to all systems. This means that two systems, or two persons, may have a different 'model of the world', i.e., different ideas of what is important as well as different knowledge and expectations. Specifically, as many have learned to their dismay, system designers and system users may have completely different ideas about how an artefact functions and how it shall be used.

The cognitive viewpoint also describes human performance as iterative or cyclic rather than as sequential. Cognition does not necessarily begin with an external event or stimulus; neither does it end with an action or response. This is in good correspondence with the perceptual cycle proposed by Neisser (1976). In CSE the perceptual cycle has been combined with the principles of the cognitive viewpoint to provide the basic cyclical model (Figure 1.7). The cyclical view explicitly recognises that meaningful human action is determined as much by the context (the task and the situation) as by the inherent characteristics of human cognition. Cognitive systems do not passively react to events; they rather actively look for information and their actions are determined by purposes and intentions as well as externally available information and events. The mistake of the sequential view is easy to understand because we are so readily hoodwinked into *observing* events and reactions and *interpreting* them using our deeply rooted model of causality. Yet an observable action does not need to have an observable event as a cause; conversely, an observable event does not necessarily lead to an observable action.

THE CLASSICAL HUMAN-MACHINE VIEW

Human information processing tended to focus on the 'inner' processes of the human mind and to describe these isolated from the work context in the tradition established by Wilhelm Wundt (cf., Hammond, 1993). This trend became stronger as time went by, partly because the proliferation of computers suddenly created a new population of users that were non-professional in the sense that they had not been specifically trained to deal with complex technology. This led to human-computer interaction as an independent field of study, which was practiced by people who had little or no experience from the processes and industries where human-machine interaction traditionally had been pursued. New generations of researchers and developers readily adopted the established mode of thinking and focused primarily on the interaction between the user and the computer with little concern for what might exist beyond that, except as an application layer. In practice this meant that there was no process over and above the human-computer interaction. The applications were in most cases driven by inputs from the user rather than having their own dynamics. Thus office and administrative applications came to dominate over process industries, power plants and aviation.

The essence of the classical human-machine view is shown in Figure 1.6, which, by using the simplest possible representation of each system as input, processing, and output functions, clearly shows the two main characteristics of the classical view. First, that the interaction is described exclusively as the exchange or transmission of input and output. Second, that humans and machines are described in the same fashion, using the finite state machine as a basis. Technically speaking, the classical human-machine view represents a closed-loop control system in the tradition of the Shannon-Weaver paradigm.

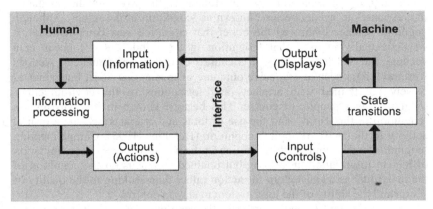

Figure 1.6: The classical human-machine view.

The Disintegrated View

The decomposed human-machine view became so accepted that the distinctive issue became interaction *with* the interface (computer) rather than interaction *through* the interface. The separation between humans and machines achieved the status of a real problem as shown by Norman's notions of the gulfs of evaluation and execution (Norman & Draper, 1986) and the very idea of usability engineering (Nielsen, 1993), and it became almost impossible to see it for what it really was – an artefact of the psychological application of the Shannon-Weaver model. Interface design became an important issue in and of itself where, for instance, the graphical user interface was seen as a problem in its own right rather than as something that played a role in how a user could interact with and control a process or application. The most recent example of this view is the notion of perceptual user interfaces (Turk & Robertson, 2000), which steadfastly continues the existing tradition, for instance, by focusing on the perceptual bandwidth of the interaction (Reeves & Nass, 2000). Yet the Shannon-Weaver communication paradigm was designed in order to deal with functioning of a communication *system*. In the context of human and machine working together we should be more interested in how the joint system performs than in how the parts communicate. While the communication between the parts and the interaction between humans and machines remain important topics, making these the dominating perspective or focus misses the more important goal of understanding how the joint system performs and how it can achieve its goals and functions.

With the gradual change in terminology from human information processing to cognition and cognitive functions, the processes that mediated the responses to events became known as 'cognition in the mind'. Although cognitive science embraced the belief that cognition was computational, it was eventually realised that 'cognition in the mind' did not occur in a vacuum, but that it took place in a context or that it was 'situated'. Actions were no longer seen as exclusive outcome of mental activities, but rather as closely intertwined with artefacts and formations of the environment – including other people, of course. This became known in its own right as 'cognition in the world', and the use of tools and artefacts became a central activity in the study of applied cognition (Hutchins, 1995). Complementing 'cognition in the mind' with 'cognition in the world' overcame one limitation of human information processing but retained the focus on the cognition of the individual as a substratum for action rather than looking to the quality of action and the ability of the joint system to stay in control.

The disintegrated view reflects the assumptions of the sequential information processing paradigm in two different ways. The first is that the predominant models for cognition in the mind are sequential or procedural prototype models (Hollnagel, 1993b). The second is that actions are seen as responses to events, mediated by internal processes and structures. This view has several specific consequences, which are acknowledged to be important for the understanding and study of humans working with technology.

- Actions are treated as a series of discrete events rather than as a continued flow of events. Yet it takes but a moment's reflection to realise that what we do always is part of one or more ongoing lines of action and that one therefore should not consider actions one by one.
- Users are seen as single individuals. In practice, however, humans rarely work alone. Humans are always involved with and depending on other humans, even though they may be removed in time and space.
- The proactive nature of actions is neglected and the focus is on response rather than on anticipation. Yet human action is more often than not based on anticipation rather than simple (or even complex) responses.
- The influence of context is indirect and mediated by input. Yet in reality we know that context has a decisive influence on what we do and how we behave, even if that influence sometimes may be hard to spell out.
- Models are structural rather than functional. For instance, information processing models focus on how information is stored and retrieved rather than on the ability to remember and recall.

CHANGING THE PARADIGM

The many problems of the disintegrated view can only partly be overcome by compensating for them via more complicated model structures and functions. Sooner or later the fundamental flaws will have to be confronted. Instead of trying to solve the specific problems one by one, the solution lies in understanding the common root of the problems, and to overcome this by proposing an alternative, integrated view. This is more than a play with words, but signifies a fundamental change in the view of how humans and technology work together. The integrated view changes the emphasis from the interaction between humans and machines to human-technology coagency, i.e., joint agency or agency in common. Agency is here used as a verb describing the state of being in action or how an end is achieved, i.e., what a system (an *agent*) does.

We have argued above that the gulfs of evaluation and execution exist only because humans and machines are considered separately, as two distinct classes or entities. While it is undeniable that we, as humans, are separate from machines, the *physical separateness* should not lead to a *functional separateness*. The physical separateness was reinforced by the Shannon-Weaver paradigm, which was developed to describe the communication. Yet for CSE it is more important to describe the functioning of the joint cognitive system, hence to join human and machine into one.

Figure 1.7 illustrates the focus on joint system performance by means of the cycle that represents how the joint cognitive system maintains control of what it does. The cyclical model is based on the ideas expressed by Neisser's description of the perceptual circle (1976), and the basic cycle of planning, action, and fact finding in the 'spiral of steps' description of purposeful action (Lewin, 1958). The model aims to describe the necessary steps in controlled performance, regardless of whether this is carried out by an artefact, a human being, a joint cognitive system, or an organisation, and it is therefore also called a contextual control model (Hollnagel, 1993b).

The cyclical model has several specific consequences for the study of how humans and machines can work together. These are considerably different from the consequences of the sequential view, and deliberately so. The net outcome is that the cyclical view offers a better basis on which to study human-technology coagency.

- Actions are seen together. The cycle emphasises that actions build on previous actions and anticipate future actions. Human behaviour is described as a coherent series of actions – a plan – rather than as a set of single responses, cf., Miller, Galanter & Pribram, 1960.
- Focus on anticipation as well as response. Since the cyclical model has no beginning and no end, any account of performance must include what

went before and what is expected to happen. The cyclical model effectively combines a feedback and a feedforward loop.

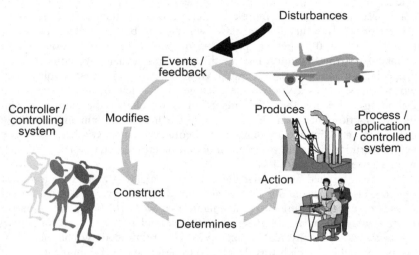

Figure 1.7: Basic cyclical model (COCOM).

- Users are seen as parts of a whole. The cyclical model focuses on coagency, on how users and environments are dynamically coupled, and on how actions and events are mutually dependent.
- Influence of situation or context is direct. In the cyclical model, context can affect the user's way of working – specifically how events are evaluated and how actions are selected, representing the fact that users may have different degrees of control over what they are doing.
- Models are functional rather than structural. The cyclical model makes minimal assumptions about components, hence about information processing. The emphasis is on performance rather than internal processes.

After the hegemony of the human information processing paradigm it may come as a surprise that the perspective on human action as being tied together, as continuous rather than discrete, is far from new. The criticism that CSE directs against the human information processing view is very similar to the criticism that functionalism directed against structuralism in the 1920s. This can be illustrated by the following quotation from a psychology textbook by H. A. Carr:

Every mental act is thus more or less directly concerned with the manipulation of experience as a means of attaining a more effective adjustment to the world. Every mental act can thus be studied from three aspects – its adaptive significance, its dependence upon previous experience, and its potential influence upon the future activity of the organism. (Carr, 1925; quoted in Schultz, 1975, p. 161)

In 1925 mainstream psychology was focused on what happened in the mind and the controversy was therefore about mental acts. At that time there were no ergonomics, no human factors, and certainly no human-machine interaction. At the present time, psychology has developed in many different directions, one of them being the practically oriented study of humans and machines. The focus is nominally on actions and human performance, but as the preceding discussion has shown it is actually still about the mental acts, although they now are called human information processes or cognitive functions. If we replace the term 'mental act' with the term 'cognitive function', Carr's criticism of structuralism can easily be applied to the information processing paradigm and cognitive science. The lesson to be learned from this is not to forget the relativism of the paradigms we rely on.

DEFINITION OF A COGNITIVE SYSTEM

CSE was formulated in the early 1980s as a proposal to overcome the limitations of the information processing paradigm that already then had become noticeable, although not yet quite obvious. A cognitive system was at that time defined as:

- being goal oriented and based on symbol manipulation;
- being adaptive and able to view a problem in more than one way; and
- operating by using knowledge about itself and the environment, hence being able to plan and modify actions based on that knowledge.

Seen in retrospect this definition clearly bears the mark of its time, which was characterised by a common enthusiasm for expert systems, knowledge representation, and quasi-formal principles for information and knowledge processing. Although the definition emphasised the importance of what a cognitive system did over detailed explanations of how it was done, it still reflected the widely accepted need to account in some way for the underlying means.

CSE was, perhaps immodestly, proposed as 'new wine in new bottles'. The new wine was the idea that the construct of a *cognitive system* could function as a new unit of analysis. This was based on the realisation that (1) a common vocabulary was needed and (2) cognition should be studied as cognition at work rather than as functions of the mind. The new bottles were the changed of focus in the domain – looking at the human-machine ensemble and the coupling between people and technology, rather than the human *plus* machine *plus* interface agglomeration. The cognitive systems synthesis arose at the intersection of what were normally boundaries or dividing lines between more traditional areas of inquiry, such as technological versus behavioural sciences, individual versus social perspectives, laboratory experiments versus field studies, design activity versus empirical investigation, and theory and models versus application and methods.

Over the years the original definition of a cognitive system has changed in two important ways. First, the emphasis on overt performance rather than covert functions has been strengthened. In the CSE terminology, it is more important to understand *what* a joint cognitive system (JCS) does and *why* it does it, than to explain *how* it does it. Second, the focus has changed from humans and machines as distinct components to the joint cognitive system. There is consequently less concern about the human-machine interaction and more about the coagency, or working together, of humans and machines. There is also less need of defining a cognitive system by itself and to worry about definitions of cognition, cognitive functions, cognitive processes, etc. Since the joint cognitive system always includes a human, there is no real need to develop a waterproof definition of cognition, or indeed to argue about whether cognition – or even intelligence – can exist in artefacts.

The revised definition of a cognitive system is *a system that can modify its behaviour on the basis of experience so as to achieve specific anti-entropic ends*. The term entropy comes from the Second Law of Thermodynamics where it is defined as the amount of energy in a system that is no longer available for doing work; in daily language it is used to mean the amount of disorder in a system. Systems that are able locally to resist the increase in entropy are called anti-entropic. In basic terms it means that they are able to maintain order in the face of disruptive influences, specifically that a cognitive system – and therefore also a joint cognitive system – is able to *control* what it does. (In cybernetics, control is defined as steering in the face of changing disturbances (Wiener, 1965; org. 1948), cf. also the discussion of the Law of Requisite Variety in Chapter 2.)

Most living organisms, and certain kinds of machines or artefacts, are cognitive systems. In particular it must be noted that organisations are cognitive systems, not just as an agglomeration of humans but by themselves. Cognitive systems appear to have a purpose, and pragmatically it makes sense to describe them in this way. In practice, the purpose of the JCS is often identical to the purpose of the human part of the system, although larger entities – such as organisations – may be seen as having purposes of their own.

As shown by the above definition, a JCS is not defined by what it *is*, but by what it *does*. Another way to characterise the focus of CSE is to note that in the case of a JCS, such as a human-machine ensemble, it is characteristic that the machine is not totally reactive and therefore not totally predictable (the same, of course, goes for humans, but this is less surprising). This lack of predictability is central for CSE. (The argument can in principle be extended to cover other combinations, such as human-human co-operation and some cases of sophisticated machine-machine co-operation.)

Consider, for instance, driving an automobile. Although seemingly a very simple thing to do, the driver-vehicle system can reasonably be seen as a JCS. First, the vehicle usually cannot be controlled by single actions but requires a combination of several actions. (This may apply even to functions that have nothing to do with driving, such as using the radio.) Second, the performance or function may not be entirely predictable, due to ambiguities in the design of the interface or because it is not clear how the buttons and controls function. Third, driving is the control of a dynamic process indeed it is steering in the cybernetic meaning of the word, which furthermore takes place in a dynamic and unpredictable environment.

The unpredictability of an artefact is generally due either to the inherent dynamics of the artefact or incomplete or insufficient knowledge of the user *vis-à-vis* the artefact – either permanently or temporarily – due to lack of training and experience, confusing interface design, unclear procedures, garbled communication, etc. In practical terms, CSE is interested in studying JCS that have one or more of the following characteristics:

- The functioning is non-trivial, which generally means that it requires more than a simple action to achieve a result or to get a response from the artefact. For more complex artefacts, proper use requires planning or scheduling.
- The functioning of the artefact is to some degree unpredictable or ambiguous for any of the reasons mentioned above.
- The artefact entails a dynamic process, which means that the pace or development of events is not user-driven. The general consequence is that time is a limited resource.

A situation of particular interest is the one where the machine controls the human – which in a manner of speaking happens whenever the user loses control. In these cases the controlled system by definition is not reactive and certainly not predictable, since users rarely do what the designers of the artefact expect them to.

The Scope of CSE

One of the motivations for the development of CSE was to provide a common set of terms by means of which the interaction between people and machines could be described. People, obviously, are natural cognitive systems, while machines in many cases can be considered as artificial cognitive systems. In 1981, as well as today, the three most important issues for CSE were: (1) how cognitive systems cope with complexity, for instance by developing an appropriate description of the situation and finding ways to reach the current goals; (2) how we can engineer joint cognitive systems, where human-machine are treated as interacting cognitive systems, and (3) how the use of artefacts can affect specific work functions. In a single term, the agenda of CSE is how we can design joint cognitive systems so that they can effectively control the situations where they have to function.

An important premise for CSE is that *all work is cognitive*. There is therefore no need to distinguish between cognitive work and non-cognitive work or to restrict cognitive work to mean the use of knowledge intentionally to realise the possibilities in a particular domain to achieve goals. Everything we do require the use of what we have 'between the ears' – with the possible exception of functions regulated by the autonomic nervous system. The cognitive content of skills becomes obvious as soon as we try to unpack them or apply them under unusual circumstances, such as walking down a staircase in total darkness. The fact that we habitually are able to do a great many things without thinking about them or paying (much) attention to them does not make them non-cognitive. It rather demonstrates that there can be different levels of control in performance. Similarly, CSE considers only the use of artefacts, without making a distinction between 'cognitive tools' and 'non-cognitive tools'. An artefact, such as a bicycle, may have been developed to support a predominantly manual function but anyone who has ever tried to teach a child to ride a bicycle will be keenly aware that this involves a very high level of cognition. There is, consequently, no requirement to have a specific discipline of cognitive design dedicated to the design 'cognitive work' and 'cognitive tools'. The engineering of cognitive systems will do nicely on its own.

Chapter 2

Evolution of Work

The human use of technology has a long history. The uptake of technology in work, and the consequent transformation of work, can be described as going through a number of stages. An understanding of the major technological innovations and solutions in the 20th century is helpful to understand how the nature of work has undergone radical changes, and to give an idea of the possibilities and problems that may lie ahead.

TECHNOLOGICAL SYSTEMS AS AMPLIFICATION

The development of technology and the use of ever more powerful technological systems can be viewed from different perspectives. Here we shall use the perspective of amplification, in the meaning of increasing something, expanding it, or making it stronger. The 'something' that is increased or made stronger is the ability – or capability – of humans to do specific things. Amplification can therefore be seen as a way of overcoming the limitations of the unaided human – whether these refer to physiological, biomechanical, perceptual, or intellectual (mental or cognitive) functions. We may tend to think that amplification is a modern phenomenon, dating mainly from the beginning of the industrial revolution, but it has been a constant facet of human endeavour from the beginning of historical time and is therefore helpful in recognising common trends and characteristics.

An early example of amplification is the invention of external representations using symbols, such as cuneiform writing from the end of the 4th millennium BC. Cuneiform writing served to improve the human ability to remember both because writing was longer lasting than individual human memory and because it increased the capacity of remembering. In that sense writing is a prosthesis or a part replacement of a human capability, cf. below. Cuneiform writing also amplified the ability to communicate over distances, although it did not increase the speed of communication *per se*, since that was limited by the ability physically to transport clay tablets from one place to another. Writing is furthermore a case of amplification of precision or

25

exactitude, since it does not depend on human recall and hence is immune to memory distortions. Writing thereby illustrates how a technology as simple as making marks in wet clay can be seen as a case of amplification.

Examples of Amplification

Other examples of amplification can easily be found by going through the history of human endeavour, especially the history of technology, as summarised by Figure 2.1.

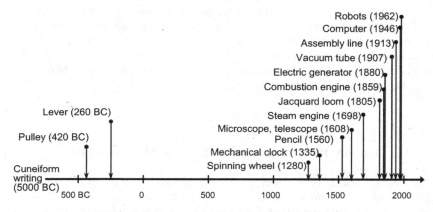

Figure 2.1: Amplification through history.

In the beginning amplification was directed at human physical abilities such as power, speed, etc. The use of animals as carriers (horses, mules, camels, oxen, elephants, etc.) amplified the speed by which a person could move, the endurance – hence the distance that could be covered – and the capacity to carry goods or materials. Another way of amplifying speed, hence making communication more efficient, was to use relay runners. This principle, which relies on a good organisation to function appropriately, was known already in Egypt around 2.000 BC, as well as in China. It reached a high point during the Roman Empire, where a postal system of messengers, known as *cursus publicus*, allowed messages to travel throughout the entire realm. It has been estimated that distances as large as 270 kilometres could be covered in a 24-hour period – a speed that was not rivalled in Europe until the 19th century! A more modern example is the Pony Express in the US, which could cover the distance of 2.900 kilometres from Missouri to California in about 10 days. This service lasted less than two years, due mainly to the completion of the transcontinental telegraph. As a means of amplifying the speed of transporting an individual rather than a message, the bicycle, the automobile and the aeroplane are other representative examples.

In terms of power, the earliest cases are linked to the discovery or invention of the basic machines, the lever, the wedge, the wheel and axle, the screwdriver (Rybczynski, 2000), and the pulley (the latter dated to 420 BC). Other classical examples are the longbow and the crossbow – or perhaps even the slingshot. (It seems as if the needs of the military frequently have been a strong driving force for inventing new means of amplification, both in the ancient world and in the modern.) All kinds of tools, such as hammers and pliers, are likewise good examples of amplification of power – as well as of precision. An artefact is generally an amplifier in more ways than one, i.e., on more than one dimension. Closer to our own time the mechanisation of work by means of windmills, waterwheels, steam engines, combustion engines, etc., exemplify amplification of power. Cranes still use the principle of the pulley combined with either hydraulic or electrical power sources, to lift objects far heavier than the unaided human is able to. The building of the pyramids, however, is ample demonstration that even the ancient people knew something about amplification, although probably by means of organisations rather than artefacts.

The amplification of power or strength often goes hand in hand with the amplification of endurance, i.e., the ability to produce steady work over long periods of time – far longer than what is possible for the unaided human. All of the power-generating machines mentioned above are examples of that. The principle of organisation of work in teams, such as the *cursus publicus*, is another way of amplifying endurance. It is also possible to amplify the precision by which work can be carried out. A simple example of that is a ruler or a pair of compasses, which enables lines and circles to be drawn more precisely than by freehand. More complex examples are various mechanical precision tools, and – more recently – manipulators and robots. The power of the machines can be used either to control the manipulation of objects that are very large or very far away, or of objects that are very small, as in the case of nanotechnology.

Whereas the amplification of the physiological output capabilities of humans is present from the earliest times, amplification of input capabilities had to await the understanding of the nature of light and development of more refined technologies, specifically the glass lens. The first useful compound microscope was constructed sometime between 1590 and 1608, followed by the invention of the telescope in 1608, both in the Netherlands. This led to the construction of the first Galilean telescope in 1609. Both the microscope and the telescope amplified the capabilities of the naked eye – the power of visual perception. The microscope made details visible that hitherto had been too small to see, while the telescope did the same for objects that were too far away or too faint for the naked eye to detect. The amplification of perception has in modern times been extended tremendously, in terms of sensory modalities and sensibility. One example is the area of remote sensing (in deep space, underwater, inside dangerous environments), another the extension of the sensibility of the human sense organs beyond the natural limits (infrared, night vision, heat sensing, the bubble chamber, etc.). At present, the marriage of computing technology and artificial sensors enable us to 'see' molecules or even atoms, even though there are interesting problems of veridicality. On a more mundane level, the presentation of measurements from a process allows the operators to 'see' what happens in the process as a prerequisite to controlling it.

More interestingly, from the CSE point of view, are the kinds of amplification that focus on human mental abilities, i.e., amplification of the mind rather than of the body. Eventually, this may lead to the amplification of intelligence outlined by Ashby (1956). We have already mentioned the invention of writing as an example, and this can clearly be seen as a way of amplifying some fundamental mental abilities – the ability to remember over extended periods of time and the ability to recall precisely. Later writing technologies (such as typeset printing) amplified the ability to communicate, i.e., the transmission of information to many recipients, as well as to utilise common (cultural) experience. Another especially interesting kind of amplification has to do with various calculating devices. Amplification of calculation has been a dream from the earliest days, and various heuristics and techniques have been used to make calculations easier. The earliest example of a calculating machine is the abacus, which is more than 5.000 years old. More recently, mechanical calculators were built in Europe in the middle of the 17th century, among others by Blaise Pascal. Since then, many inventors and philosophers have tried their hand at the invention of calculating machines. Charles Babbage aside, it was not until the technology of electrical circuits became practical that really powerful calculating machines could be built – analogue calculators and integrators notwithstanding (Goldstine, 1972). The invention of the digital computer, which initially was used for military purposes such as cracking the Enigma code or calculating ballistic trajectories, not only enabled calculation on an unprecedented scale but also brought about the amplification of logic and reasoning.

Amplification of Control

For CSE, the most interesting type of amplification concerns the ability to control a dynamic system or process. Maintaining control of complex systems requires the ability to counter the effects of disturbances and noise so that the system can continue to function and remain within operating or safety limits. A classical example of that is the development of the marine chronometer (Sobel, 1998). Control is thus intrinsically related to the concept of a dynamic process: if nothing changes, then there is nothing to control. Everything, however, changes over time, although at different time scales. In the context of CSE we are interested in the control of systems that change noticeably over the time scale of human activity, ranging from seconds to hours – or perhaps even days – depending on the task at hand.

One type of control is feedback-based, or error-based, control, as in the closed-loop system shown in Figure 2.2. Despite the modern sound of the name the principle was known as far back as the second century BC, where Ktesibios in Alexandria invented a self-regulating flow valve. In joint cognitive systems the controller is often a human operator rather than a technological device. In order to perform effectively as a closed-loop controller, the operator must continuously determine the state of the process by reading the value of the appropriate variables. If any disturbances or deviations are detected, the operator must provide corrective input to the process by selecting and executing the actions necessary to counteract or compensate for the disturbances. Following that, the effects of the corrective input must be determined, to see if the response was satisfactory, and so on. To perform such continuous control can be very demanding, specifically if the time constants are small, i.e., if the process changes rapidly. (The change can be due to external influences, but also to insufficient reliability of the process or machine itself, e.g., inadequate workmanship or quality. That was often the case in the beginning of the industrial revolution, when the first controllers were developed.) There is therefore an obvious need to amplify the human ability to control, which basically means automating the detection-regulation loop. Relieving the operator from the need closely to monitor the process not only improves the efficiency of the closed-loop control but also enhances the ability to control on higher levels. The amplification of control thus exemplifies the phenomenon of replacement or substitution of a human function by an artefact.

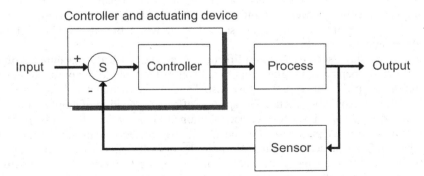

Figure 2.2: Basic feedback control system.

In the 20th century we have witnessed a proliferation of controllers and a significant development in control techniques. In the industrialised societies we are in our daily life surrounded by numerous controllers and servomechanisms, such as all the household equipment, cars, machines, computers, etc. In more recent times the advent of AI (Artificial Intelligence) has led to new types of controllers where the classical feedback-loop has been augmented by complex logic.

The amplification of control is thus both the replacement of human control and the development of controlling artefacts that go beyond what was ever humanly possible. Control is obviously closely linked to the issues of automation, which will be discussed in detail in Chapter 6. In a sense, amplifying or augmenting control on one level therefore means losing control on another. This happens when a function is taken over by an automated controller, which may or may not inform the human controller of what it is doing – most frequently the latter (Sheridan, 2002). There may in fact be good reasons not to inform the operator, since that would produce a considerable amount of potentially unnecessary information. There are, however, situations where the missing information can be crucial. This is a perennial problem for automation design, which shall be discussed later on.

Effects of Amplification

As shown by Figure 2.1, the invention of artefacts to amplify human capabilities is unevenly distributed over time. Although some artefacts go back thousands of years, the pace did not begin to pick up until the industrial revolution towards the end of the 18th century. Since then there has been a steady stream of inventions, of which but a few are included in Figure 2.1, with a second spur of growth from around the middle of the 20th century. This was due to several major conceptual developments in the meta-technical sciences that led to, among other things, the digital computer. The most important of these have already been discussed in Chapter 1. They had important consequences not only for the technology that constitutes the working environment, hence the substance of work, but also for thinking about humans at work.

As we have seen above, amplification can be described on a number of different dimensions, ranging from the power of the body to the power of the mind. Amplification can therefore also have many different effects, some harmful and some beneficial. The specific effects of the kind of amplification that is associated with extensive use of automation, as well as the problem of artefacts seen as prostheses or tools, will be treated extensively later in Chapter 5. At present we will only note the overall effects of amplification with regard to the balance between manual and mental work. This is shown in Figure 2.3 as the changes in demands to muscle power and brainpower, respectively. Needless to say, the two curves have been added to illustrate general trends and to clarify our argument and do not represent empirical data.

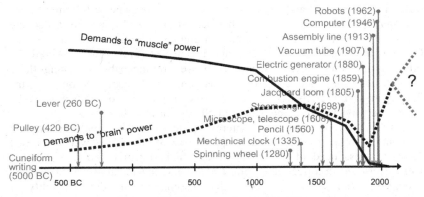

Figure 2.3: Effects of amplification.

One distinctive effect of amplification is a reduced demand on muscle power, i.e., on the physical strength and capabilities of humans. The effect is discernible from antiquity and was, indeed, one of the main motivations for this kind of amplification. The trend, as illustrated by Figure 2.3, has been a slow but steady reduction until the beginning of the industrial revolution. Mechanisation and the advent of power generating machines, such as the steam engine or the internal combustion engine, meant that human muscle power could be amplified almost infinitely, as long as the machines could be controlled. A further significant drop came about with the invention of the electrical motor, and even further with the amplification of control such as servomechanisms and automation. In that sense an industrial robot represents the ultimate solution since it completely takes over manual work from humans – for instance in an assembly line. Only a very small physical effort is needed to control the robot, but a higher mental effort may be required to program and control it.

Another distinctive effect of amplification is the changing demands to brainpower, in particular to the functions of monitoring and planning. As indicated in Figure 2.3 this development is not nearly as simple as for the amplification of muscle power. To begin with, the demands on brainpower grew as the demands on muscle power were reduced. This happened because there was no way to control the mechanised artefacts, except by the 'naked' human mind. It was not until the James Watt's invention of the flying-ball governor, and perhaps more significantly the invention of the punched cards controlling the Jacquard loom, that demands on brainpower started to go down. Further developments in technology and electronics, as well as developments in control theory, have dramatically reduced these demands. Yet as Figure 2.3 indicates, it is uncertain whether this development will continue. At present we seem to be facing a situation where the power of technology almost paradoxically has led to an increased need of brainpower – in many cases because technology has been used in awkward and inadequate ways, such as illustrated by the notion of clumsy automation (Wiener, 1988). Despite the potential for the amplification of brainpower offered by information technology, it seems as if we are incapable of using these possibilities constructively. This was one significant reason for the development of CSE as described in Chapter 1 and will consequently also be one of the ways in which the value of CSE will have to be proved.

As a final comment, it is interesting to note that amplification and attenuation go hand in hand. The amplification of physical power meant that there was less need of bodily strength, which in combination with other factors resulted in a decline of physical fitness. In the same way it may be expected – and perhaps feared – that automation and the possibility of amplifying cognition may lead to an atrophy of mental powers and the ability to think. This was pointed out by Ihde (1979), who stated that any tool both amplifies a function and reduces another. The loss of physical or cognitive skills will inevitably lead to a dependence on machines, which on the whole makes our systems more vulnerable.

AMPLIFICATION AND INTERPRETATION

In the relationship between humans and artefacts, the latter can serve as either a tool or a prosthesis. One prerequisite for technology to be used in a productive way is a well-formulated view of what human-machine systems are and a principled position to what the role of technology should be *vis-à-vis* the operator. Taken at face value the term human-machine interaction simply means the interaction between a human (an operator) and a machine or an artefact (a process and/or a computer) with the implied purpose of carrying out a certain task or achieving a certain goal. One thing to note is that the artefact usually serves as a kind of intermediary, and that the interaction therefore is *through* the artefact rather than *with* the artefact. This is characteristically the case for the control of dynamic processes, particularly in systems where the interaction is mediated by computers and information technology. In such cases the artefact is not itself the target of the interaction but rather the tool or mediator by means of which the task can be accomplished. The operator therefore interacts with a process through an artefact.

The artefact as an intermediary can have two completely different roles (Figure 2.4). In one it serves as an extension of the body and the mind, hence as an amplifier as discussed above. The artefact may provide process data to the operator (as well as control data to the process) and in doing so amplify selected cognitive functions, e.g., discrimination or interpretation. (While we are here concerned mainly about the cognitive functions, amplification often addresses a mixture of manual and cognitive capabilities as noted above.) The amplification highlights those aspects of the experience that are germane to the task and simultaneously reduces or excludes others, controlled by the operator. In this role the artefact serves in an *embodiment* relation, defined as a relation where the artefact is partially transparent, and thus part of the operator rather than of the application (Ihde, 1979).

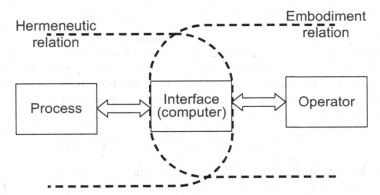

Figure 2.4: Embodiment and hermeneutic relations.

In the second role, the artefact serves as an interpreter for the operator and effectively takes care of all communication between the operator and the process. In this role the artefact is partially opaque and therefore part of the process rather than of the operator. Put differently, the operator has moved from an experience of the *world* through the artefact to an experience of the *artefact* embedding the world. It is thus the state of the process as represented by the artefact – especially when that is a computer – which assumes importance. In the extreme case there is no actual experience of the process except as it is provided by the artefact. In this role the artefact serves in a *hermeneutic* relation to the operator and functions without operator control.

Tools and Prostheses

The roles of the artefact as an amplifier (tool) and as an interpreter (prosthesis) are clearly distinguishable. In industrial process control, the operator is often removed from the process and may not even have any experience with it outside of the control room. The representation of the process in the control room, through computerised display systems, therefore becomes the real process. The operator's understanding may accordingly come to depend on the surface presentation of the process and the operator's experience may be representation-dependent shallow knowledge (Gallanti et al., 1988). This suggests the possibility that the more 'advanced' the HMI is and the more sophisticated the computer's functions are, the more likely we are to depend on it, hence to use the computer as an interpreter rather than as an amplifier. In the embodiment relation the machine shows a partial transparency that becomes part of the experience, whereas in the hermeneutic relation the machine creates an opacity that becomes objectified, hence the 'object of experience'. The distinction between tools and prosthesis reflects one of the primary concerns of CSE and will be discussed extensively in following chapters.

The two roles are not mutually exclusive but rather represent two different ways of viewing a joint cognitive system, which closely resembles the distinction between using artefacts as tools or as prostheses (Reason, 1988). As an example of technological developments, artificial intelligence (AI) makes it possible to amplify the operator's cognitive functions, hence to serve as a tool. But at the same time AI may be used to enhance the artefact as an interpreter, thereby strengthening its role as a prosthesis. This may happen as a simple consequence of the increasing complexity of the processes and the perceived need to provide access to compiled expertise. Thus the very development of expert systems may unknowingly favour the computing artefact's role as an interpreter (Weir & Alty, 1989). The development of operator support systems and advanced HMI functions should, however, rather aim at amplifying the capacities of the operator.

A SHORT HISTORY OF HUMAN-MACHINE INTERACTION

The problem of using machines effectively is as old as technology itself and the concern for what we generally know as human-machine interaction dates from antiquity. For many centuries the primary problem was how to build ever more powerful artefacts, rather than how to interact with them. Human-machine interaction itself became a problem only recently – historically speaking – as a result of a burst of technological innovation. In a characterisation of the 20th century it has been remarked that the first fifty years were mainly spent on harvesting the fruit of inventions made in the 19th century in physics, engineering, biology, and medicine. In contrast to that, the second half of the 20th century was characterised by a flood of inventions and technological innovations – although most of them were based on earlier theoretical developments, particularly from the 1930s and 1940s.

One result, already discussed in Chapter 1, was the increasing power and complexity of technological systems. Since the new conditions for work were predicated on what technology could do rather than on what humans needed, the inevitable result was that the human became the bottleneck of the system. The so-called shortcomings of human performance in human-machine systems became conspicuous and created a need to design machines, operations, and work environments to match human capabilities and limitations (Swain, 1990). Although this need had gradually been growing in the production industry, it was accentuated by the technology burst that came with the Second World War (1939-1945). The observed shortcomings were, of course, not due to a sudden deterioration of human capabilities, but rather due to changing demands from increasingly complex technological systems. The demand-capacity gap created in this manner was discussed in Chapter 1.

In a wider perspective, the changes in the relations between humans and machines can be attributed to four technological advances, which materially shaped the nature of the questions facing system designers (Figure 2.5). The advances can be seen as a shorthand rendering of the history of process control, following a suggestion by Kragt (1983).

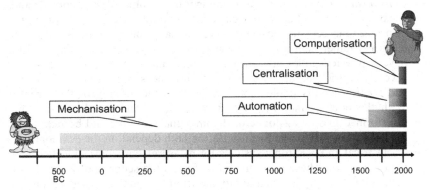

Figure 2.5: Four stages in the development of work.

- The first advance was *mechanisation*, the widespread application of mechanical power to amplify or replace human muscle power. Although mechanisation is primarily associated with industrial work, specifically the developments leading to the industrial revolution, it can be found in many other areas as well, such as agriculture, the military and domestic tasks. Mechanisation increased the efficiency of individual work but also led to a specialisation of activities. It brought the concept of tasks and task elements to the fore, as epitomised by the principles of scientific management (Taylor, 1911). The process of mechanisation is still ongoing, as we invent artefacts to take over new tasks and functions. As a result, it is possible in the industrialised society to go through the day without spending any substantial physical effort, except perhaps walking to and from the car.

- Mechanisation was followed by *automation*, which put elemental processes under the control of an automatic system rather than of a human worker. Simple regulators existed long before the invention of electrical and electromechanical devises (from the self-regulating flow valve of Ktesibios to Watt's flying-ball governor), but it was not until the appearance of mechanised logic and electronic components that automation gained speed. This development started slowly in the 1930s but soon picked up speed and has continued ever since. (Although the word *automation* was first used in 1948 to describe the automatic control of the manufacture of a product through a number of successive stages, the word *automatic* goes back, in English, to 1748.) Automation generated a concern for the proper specification and representation of process information and for the development of control systems, particularly when it was reinforced by centralisation.

- *Centralisation* meant that a large number of (automated) components and subsystems could be controlled from a single location. Large-scale centralisation became possible when signals could be transmitted by electrical rather than physical means. Before that happened, the need to have a physical link between, e.g., a tank and the control room would severely restrict the distance from which a measurement could be made and control actions implemented, as well as the validity of the signals. With the new technology, measurements made in one place could be transformed into an analogue or digital signal, which could be read from far away. Automation and centralisation both depend on the development of electrical and electronic systems for measurement, communication and remote control. These have greatly increased the scope and flexibility of methods for getting information to and from humans in system operation. The possibility of arbitrary relocation and centralising of operating points have fundamentally changed the human-machine communication problem. In parallel, the display and control facilities available to the designer have been greatly extended.

- *Computerisation*, finally, enabled a significant expansion of automation as well as of centralisation. In addition to increasing the speed and complexity of information processing, computerisation enabled the use of very large data bases in both off-line and on-line conditions. It also increased the flexibility of design of interfaces (information presentation) and control devices, and it made it possible to propose and develop sophisticated support functions. Unfortunately, the extensive use of computers created an equally large number of possibilities for making simple tasks unnecessarily complex and for inadvertently introducing task requirements, which were clearly beyond human capacity.

Computerisation differs from the earlier changes of mechanisation, automation and centralisation because it, in a sense, bears the answer to its own problem. By this we mean that the problems that are created by the widespread use of computers, such as information overflow and the fragmentation of work, to a considerable extent can be solved by using computers and information technology to improve the design of the working conditions themselves. On the one hand, the use of computers in processes has increased the requirements to operators. On the other, the capacity of operators can conceivably be increased by use of computers and artificial intelligence in support systems. It may, however, not be the best way to solve the problem.

The Conspicuousness of Cognition

Changes to the nature of work, from doing to thinking, has made cognition more conspicuous. The importance of cognition, regardless of how it is defined, as a necessary part of work has grown after the industrial revolution. Cognition is, unfortunately, also necessary to cope with the dilemmas, double binds, and trade-offs that occur as by-products of a myopic use of technology to further technical capabilities. The constant striving for improved efficiency often leads to multiple and sometimes inconsistent goals, organisational pressures, and clumsy technological solutions. The contemporary work environment can be characterised as follows:

- Cognition is distributed rather than isolated in the mind of a thoughtful individual, and cooperation and coordination are ubiquitous. Operators are embedded in larger groups and organisations, which together define the conditions for work, the constraints and demands as well as resources.
- People do not passively accept technological artefacts or their general conditions of their work but actively and continuously adapt their tools and activities to respond to irregularities and disturbances, and to meet new demands.
- Technological development is rampant, leading to more data, more options, modes, and displays, more partially autonomous machine agents, and inevitably to greater operational complexity (cf. Chapter 1).
- Technology is often used in ways that are not well adapted to the needs of the operator. This leads to new demands that tend to congregate at the higher tempo or higher criticality periods of activity. As the complexity of

systems grows it manifests itself as an apparent epidemic of failures labelled 'human error'.

The increased conspicuousness of cognition is a reminder that the nature of work has changed significantly. Work has always been cognitive, but cognition is more important now than it was before. Cognition is, however, not just another element that can be added to a model, or another category of decomposition. The importance of cognition does not mean that we must focus on cognition as a separate process or function either 'in the mind' or 'in the world'. It rather means that we must bear in mind that cognition signifies the ability of humans to maintain control of their environment, hence of their work.

Changing Balance between Doing and Thinking

Using a distinction between working with the body and working with the mind – or between *doing* and *thinking* – the long term change to the nature of work has been a reduction in work with the body and an increase in work with the mind (cf. Figure 2.3). There is now less direct, closed-loop interaction (control and regulation) and more open loop interaction (monitoring and goal setting) than before – both in a relative and an absolute sense.

The change in relations between the two types of work can be illustrated as shown in Figure 2.6. Although the curve is fictive rather than empirical, two features are reasonably correct. The first is that the balance changed rapidly around the middle of the 20th century, for reasons described in Chapter 1. The second is that the current situation is one of very little manual work and more and more cognitive work, i.e., work with the mind. The rate of change is levelling off for obvious reasons – a saturation point is reached as work with the body mostly has been taken over by machines. The nature of work will, however, continue to change and technology will gradually make inroads on the more complex forms of work with the mind. In order to be ready for that we must change the perspective from that of human-machine interaction to that of the functioning of joint cognitive systems.

Loss of Work-specific Information

One outcome of the changing nature of work is that work-specific information gradually is lost. All working environments are different as long as one is at the physical place of work: at the assembly line, in the refinery, at the furnace, etc. Here the process can be seen, heard, smelled, felt, and so on. However, as people become removed from the actual working environment to a sterile control room, they get only the information that is transmitted to them, which is but a subset of the information that exists in the working environment. The information transmitted to the control room is limited both by the capabilities of sensors and channels, and by the designers' perception of operator needs (cf. Chapter 9). This outcome is reinforced by the extended use of automation. Processes may be very different, but automation and control loops are very similar; indeed, that is one of the reasons why automation is so powerful. Yet as the information presented becomes information about the control loop rather than about the process, work-specific information gets lost.

This development is reinforced by attempts to standardise human-machine interfaces, and by the use of common presentation technologies. As soon as information is found on a visual display unit (VDU) rather than on individual instruments placed on walls and panels, it loses many of the specific characteristics. Before the use of workstations, information presentation was determined by industry practices in the domain, by the availability and suitability of signals and measurements, and by the physical characteristics of the place of work. After the introduction of workstation technology, information presentation had to conform to the limitations of the new medium in terms of colours, shapes, contours, details of resolution, dynamics, etc. Spatial information about the location of measurements was lost because the process was seen through a narrow window rather than on a large \surface such as a wall. Characters (numbers and letters), colours, shapes, and forms also became standardised – or at least affected by common constraints. The differences in processes were therefore represented by the contents of the information, rather than by the structure (form) and localisation of the information.

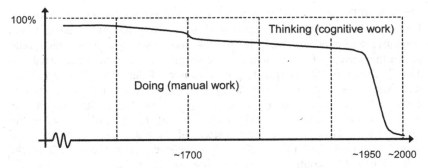

Figure 2.6: Changing balance between manual and cognitive functions.

The gradual elimination of work-specific information makes it more difficult to achieve a satisfactory performance of the tasks. One reason is that the people charged with controlling the system would no longer have the necessary practical experience with the physical process, but be proficient only in manipulating it via the interface. By being confined to working with symbols on the screen they easily lose a feeling for what the process is about. Starting a main circulation pump is never as easy and effortless as clicking a mouse on a screen icon. A physical pump can be a major piece of equipment that may require several minutes to get into operation, yet the representation by a pump symbol completely fails to provide that information. As long as operators retain the practical experience from working in the process, they may be able to understand that the icon represents a physical pump with specific characteristics. But if they only have practiced controlling the process from the control room or a simulator – perhaps with a quick walk-through of the plant thrown in for good measure – they will lack information that may be essential in abnormal conditions, such as accidents. A second reason is that the organisation of information is constrained by the characteristics of the medium rather than by the characteristics of the process. It consequently becomes more difficult to operate the process, to find the information needed, and to carry out control actions. When many devices are operated in the same manner, the likelihood of making a mistake goes up and the facility by which a device can be controlled goes down. Standardisation has its drawbacks, and in the case of control environments one should be careful about taking away information just because it does not meet the constraints of the new media.

THE LAW OF REQUISITE VARIETY

The Law of Requisite Variety was formulated in cybernetics in the 1940s and 1950s (Ashby, 1956). It is concerned with the problem of control and regulation and expresses the principle that the variety of a controller should match the variety of the system to be controlled, where the latter usually is described in terms of a process plus a source of disturbance. The Law of Requisite Variety states that the variety of the outcomes (of a system) can be decreased only by increasing the variety in the controller of that system. The purpose of the regulator is to keep the variety of the system's output within certain limits. This decrease in the variety of the outcomes can be achieved only if the variety of the regulator is at least equal to the variety of the system. Effective control is therefore impossible if the controller has less variety than the system. This means that the regulator must *be* a model of the system in the sense that the regulator's actions are isomorphic to the system's actions as seen through a mapping *h* (Conant & Ashby, 1970, p. 95). In an expression that achieved a certain degree of fame, the authors expressed the principle as 'every good regulator of a system must be a model of that system' (Conant & Ashby, 1970).

The assertion of Conant & Ashby that a regulator must *be* a model of the system was by many interpreted to mean that the regulator must *have* a model of the system although this is not the same thing. For instance, Pew & Barron (1982) argued that 'a model of human performance implies the existence of a model of the environment or system in which that performance takes place'. This meant that the model was seen as an explicit form of knowledge representation rather than as a concept. In retrospect this reformulation of the assertion missed an important point expressed by the Law of Requisite Variety, namely that the model is the basis for generating control input to a process in order to keep the variety within given limits (Hollnagel, 1999a). The model is therefore successful or adequate only if the predicted and actual performances are isomorphic, which means that the criterion is *functional adequacy* rather than *structural similarity*. The requisite variety should refer to the performance of the joint cognitive system as it can be observed from experience and empirical studies – but not from theoretical studies.

It makes obvious sense that a model needs only enough variety to account for the observed variety. There is no good reason to account for hypothetical functions, such as many of the human information processing models do, not least because such functions can be only inferred and not observed. The decision about how much variety is needed can be made by applying the Law of Requisite Variety pragmatically: if the predicted performance matches the actual performance sufficiently well, then the model has sufficient variety. This furthermore relieves us of the affliction of model validation. The catch, of course, lies in the meaning of 'sufficiently' which, in any given context and for any given purpose, must be replaced by a well-defined expression.

Models of the Human as Controller

As described in Chapter 1, the sequential human information-processing model had by the end of the 1970s become the *de facto* standard for human-machine systems and human-computer interaction research. This all-absorbing concern for the internal, mental model unfortunately led to a neglect of other aspects, of which the most important was the flexibility and variability of human performance. Although this was formally recognised by several of the information processing models, none of them were able to represent that important feature (cf. Hollnagel, 1983; Lind, 1991).

Within these fields of research the problem of how operators controlled a dynamic process was usually replaced by the problem of how to design the human-machine interface. This change reflected the control engineering perspective, which traditionally started by decomposing the system into its various parts and then went on to describe each individually in terms of inputs, outputs, and internal state transitions using the finite state machine as an archetype. The roots of this seductively simple approach go back at least to the persuasive arguments of Craik (1947 & 1948). The object of study was in this way transformed into smaller and more tractable problems of transmitting information and control actions across the interface, and especially into problems about the operator's cognitive processes and mental models.

This kind of description is, however, inappropriate since humans definitely are not optimal controllers or finite state machines – notwithstanding the efforts to turn them into that, e.g., by strict training regimes. First, human performance is based on feedforward as well as on feedback. Feedforward control is more powerful than feedback control, as correctly pointed out by Conant & Ashby (1970, p. 92). In spite of that, practically all information processing operator models represent feedback or error-based control only. Second, human performance is influenced by the conditions under which it takes place, and control can apparently occur on different levels with consequences for the quality of performance. The transition between levels is itself an important characteristic of human performance, which generally is adaptive as shown, e.g., by the early studies of responses to information input overload (Miller, 1960). Both of these characteristics set humans apart from technological artefacts, and their modelling therefore requires an approach that is completely different from the IPS tradition.

The Joint Cognitive System (JCS)

The importance of considering the characteristics of the human as a controller is central to the view of CSE. In consequence of that, the focus of investigation and analysis is the JCS rather than human-machine interaction. This difference is not just semantic but also pragmatic. In terms of semantics, 'human-machine interaction' or 'human-machine system' introduces a distinction between humans as one part and machines as the other, which makes the interaction between them an essential mediating process. By doing so, the view of the human-machine system as a whole is pushed into the background and may even be completely lost. Yet control is accomplished by the JCS and depends on human-machine coagency rather than on human-machine interaction. This becomes immediately obvious when we consider the performance of a group of people, such as a team of operators or an organisational unit. Here it is the performance of the group that counts, rather than the performance of the individuals, and the co-operation and congruence of system 'components', i.e., individuals, is important (Hollnagel, 1999b). Yet the same line of reasoning can equally well be applied to systems where the 'components' are a mixture of humans and machines. The decomposition view must therefore be complemented by a view of coagency. The focus should consequently not only be on cognition and the internal mechanisms and processes of the components, but also on how they interact and co-operate.

Control and Cognition

The study of the human controller was from the very start a study of the internal functions of the controller, either as information processes or as cognition. In view of the above considerations it is, however, reasonable to consider whether control defines cognition or *vice versa*. This is not meant to be a question of whether humans *have* cognition, since they obviously do by any reasonable definition of the term. It is rather a question of whether it is necessary to study human cognition in order to study the performance of human controllers, and specifically whether a model of a controller *must* be a model of cognition.

A cognitive system is defined by its ability to modify its pattern of behaviour on the basis of past experience in order to achieve specific anti-entropic ends. Given this definition, one issue is clearly whether it is cognition that provides the ability to control, i.e., whether humans are able to control what they do (plan, understand, etc.) because they have cognition. If that is the case, it leaves the question of what the working principles are of the automated system or artefacts that are designed as controllers. Clearly, it cannot be cognition in the same sense as humans have it. Yet there must be some common quality, since they are able to function effectively as controllers. If, on the other hand, the starting point is a system's ability to control, that is to be a cognitive system in the above meaning of the term, then the question becomes whether one could define cognition by the ability to control. In such a case, humans would have cognition because they can control, and they would have very sophisticated cognition because they are able to do many other things as well. Artefacts that can control, at least according to the definition above, would also have cognition although in a more limited sense. The same would go for organisations, animals and other living organisms.

The consequence of this argument is that there is no *a priori* reason to assume that a model of the human controller must be a model of information processing or of cognition in the psychological sense. On the contrary, if the starting point of a model is a complicated theory of human information processing, the outcome will inevitably reflect the theoretical variety rather than the observed variety. This line of reasoning is far from new, and has been put forward by leading researchers such as Broadbent (1980), Marr (1977) and Neisser (1976). Broadbent specifically warned that

... one should (not) start with a model of man and then investigate those areas in which the model predicts particular results. I believe one should start from practical problems, which at any one time will point us towards some part of human life. (Broadbent, 1980, p. 117)

Ulrich Neisser, who must be considered as one of the founding fathers of cognitive psychology, early on warned against psychology committing itself too thoroughly to the information processing model and argued for a more realistic turn in the study of cognition (Neisser, 1976). This led to the following four requirements for cognitive psychology:

- Firstly, a greater effort should be made to understand cognition out of the laboratory, i.e., in the ordinary environment and as part of natural purposeful activity.
- Secondly, more attention should be paid to the details of the real world and the fine grain structure of the information that is available.
- Thirdly, psychology should come to terms with the complexity of cognitive skills that people are able to acquire and the way in which they develop.
- Finally, the implications of cognitive psychology for more fundamental questions of human nature should be examined, rather than being left to "behaviorists and psychoanalysts."

Neisser was reflecting on the then burgeoning discipline of experimental cognitive psychology, which had yet to spawn cognitive science. Even at that time the preponderance of mental models was obvious and Neisser sternly warned that

(w)e may have been lavishing too much effort on hypothetical models of the mind and not enough on analyzing the environment that the mind has been shaped to meet. (Neisser, 1976, p. 8)

As we all know, this warning was not taken seriously. If anything, the attraction of mental models grew during the 1980s, spreading from cognitive psychology into many other disciplines (cf. Hollnagel, 1988). It was not until the early 1990s that the need to consider cognition as a part of action, rather than as an internal process, was generally accepted. From a modest beginning (Suchman, 1987) this developed into a major school represented by such works as Klein et al. (1993) and Hutchins (1995). CSE had also from the beginning emphasised that the focus should be on problems that are *representative* of human performance, i.e., that constitute the core of the observed variety.

Disjoint and Joint Systems

Earlier in this chapter the concepts of the embodiment and hermeneutic relations were introduced as a way of distinguishing between different types of amplification. The principle of the Law of Requisite Variety provides a supplementary perspective on these relations, which can help to understand the global picture.

- The hermeneutic relation characterises a situation where the artefact serves as an interpreter for the operator and effectively takes care of all communication between the operator and the process. In this role the artefact must be considered as part of the process rather than of the operator. The operator now has to cope with the variety of the process as well as the variety of the artefact – which effectively becomes a pointless amplifier of the variety of the process. The hermeneutic relation therefore increases the demands to the operator who has to cope with the complexity of the control system as well as the complexity of the process.
- The embodiment relation characterises a situation where the artefact serves as an extension of the body and as an amplifier of the operator's capabilities. Here the artefact helps to exchange information between the operator and the process (status information to the operator and control information to the process) while at the same time facilitating essential control functions such as discrimination or interpretation. The requisite variety of the operator as a control system is therefore supplemented by the variety of the artefact, which means that the joint cognitive system is better able to control the process.

Despite the disallowance of the IPS view there is no problem in referring to the exchange of information between the artefact and the operator. Using such terms does not commit us to embrace the disjoint system view that is inherent in the information processing approach. The consequence is only that we should be careful about defining what the entities of the joint cognitive system are and where the boundary between the joint cognitive system and the environment lies. There will always be a transmission of information – or mass and energy– across the boundary, as well as between the entities that make up the system (at least until a better paradigm comes along). But the delineation of the necessary system structures should be based on an appreciation of the essential system functions, and not 'accidental' physical differences.

Amplifying the Ability to Control

A JCS is defined in terms of its ability to maintain a dynamic equilibrium, thereby increasing the chances of continued functioning – and, in a sense, of survival – in an unpredictable environment.

The term 'intelligent' has often been applied to the use of artefacts to ornament otherwise lacklustre technological functions. The possibility of developing an intelligent artefact has had, and continues to have, a strange attraction on people, including otherwise sensible men of letters. It is nevertheless both easier and less pretentious to consider the developing artefacts that can maintain control of a situation, even if this is couched in terms of purposiveness (e.g., Rosenblueth, Wiener & Bigelow, 1968; org. 1943). One advantage of cybernetics was that it enabled the building of machines that could behave as if they had a purpose, thereby bypassing much of the discussion about intelligence.

For CSE the ability to maintain equilibrium or to be in control is paramount. This in turn requires the ability to make predictions or to use feedforward control. From a practical point of view, the ability to predict involves three essential functions, namely (1) the generation of plausible alternatives (of future developments), (2) the characterisation of alternatives *vis-à-vis* current objectives, and (3) the evaluation and choice of most appropriate alternatives, given performance conditions (constraints).

Information technology can support generation of alternatives by supplementing the human ability for finding innovative solutions with the machine's single-mindedness and speed. Human creativity notwithstanding, it is often necessary to guard against biases in judgement and in taking short-cuts in finding the proper alternatives. We know that assumptions and expectations can easily lead operators astray. Machines are single-minded and stubborn, but since they are fast this characteristic can be used to their advantage. (It is this character that makes it possible for a machine to play good chess, not the ability to be creative or have insights.) The use of artefacts may therefore amplify generation of plausible alternatives.

When it comes to the characterisation of alternatives, machines are less efficient. The characterisation of alternatives is not so much a question of computational speed, as of finding appropriate criteria. Human decision making is powerful – and fragile – because of the ability to develop *ad hoc* criteria. Machines can possibly check whether the criteria are consistent and unambiguous but only in a local sense (Silverman, 1992) referring to the set of alternatives being considered. The characterisation of alternatives is conceivably something that could benefit from a JCS approach.

Finally, the evaluation and choice of a best alternative is as yet something that is beyond the practical capabilities of artefacts, hence best left for humans to do. The generation of alternatives, their characterisation, and finally the choice have each been the subject of much research in AI and human factors engineering. Yet it is only by considering them together, rather than as separate functions, that they make sense. This perspective is offered by the notions of a JCS and coagency.

The focus on the ability to control has another consequence, namely that the search for solutions is guided by what the JCS needs to do, rather than by how it is built and what it is built of. In the structural approach, the starting point often constrains the development process, and thereby the results. According to a Japanese proverb, 'if your only tool is a hammer, all your problems will look like nails'. The availability of powerful tools and detailed models of the structure of cognitive agents directed the attention towards the structural aspects of the artefacts. In contrast to that, CSE emphasises the functional aspects, i.e., *what* a system does rather than *how* it does it. It is the purpose of CSE to provide a functional solution to amplification, so that the result represents an embodiment rather than a hermeneutic relation.

Chapter 3

The Basics of a Science

It is a paramount characteristic of a scientific study that it is systematic. It must be based on a model, from which can be derived a classification scheme and a set of methods. For measurements, it is important that they not be based on loosely formulated folk models. This chapter describes the basic principles behind CSE, as the study of how cognitive systems perform rather than of cognition as a mental process, and introduces the three main threads: coping with complexity, use of artefacts, and joint cognitive systems.

MODEL – CLASSIFICATION – METHOD

The purpose of science is to produce valid, scientific knowledge about a particular domain or problem area. Put differently, the purpose is to enable coherent and systematic descriptions of a set of observable phenomena. The observations can be unaided in the sense that they are carried out by humans without artefacts to support the observation itself, although some kind of recording often is used to avoid the dependence on human remembering. An example of unaided observation is *in situ* studies of people at work. The observations can also be aided or assisted, meaning that specific tools or artefacts are used either to improve the resolution or endurance of human senses, or to extend perception beyond the natural range of human receptors. An example of that is the cloud chamber used in particle physics or computer-coloured NMR scans. Different ways of making observations by themselves raise some fundamental epistemological issues, which we however shall refrain from going into here.

Scientific knowledge and scientific descriptions entail the use of a set of categories that describe the phenomena being studied and a set of methods or rules that prescribe how the categories should be applied and define the difference between correct and incorrect uses. The descriptions can either be aimed at *understanding* what has happened, in particular by building causal explanations, or *predicting* what may happen in the future, for instance when constructing new artefacts (e.g. Petroski, 1994; Sample, 2000) or designing new tasks. In both cases a body of valid scientific knowledge is necessary for explanations and predictions to be made, regardless of the nature of the domain.

The need to have the three elements of model, classification scheme, and method is not specific to CSE but can be found across the whole spectrum of science as, for instance, in classical botany or particle physics. The latter also provides a good example of the dependency between the model and the classification scheme. The changes in the understanding of the nature of the fundamental particles – from electron to quarks – have led to changes in the categories for classification of the observed phenomena, as well as to changes in the methods. Since a cognitive system is defined by its ability to maintain control, the three elements must therefore meet the following requirements.

- The model must be about how a JCS can maintain control in a dynamic environment. Although cognition is an essential part of human action, CSE does not need models of cognition *per se*, but rather models of coagency that correspond to the requisite variety of performance. The existence of the model is a useful reminder that we are not dealing with an objective reality. This is particularly important in the field of behavioural sciences where the origin of observed differences may easily be lost in exegeses.

- The classification scheme is an ordered set of categories needed to define the data and knowledge that are used by CSE. In the words of Sherlock Holmes, "it is a capital mistake to theorize before one has data". Yet it is an even greater mistake to believe that data can be collected without some kind of theory or concept. Data exist only in relation to a set of concepts or a classification scheme and are not just waiting to be picked up by a meandering scientist. In CSE the categories should be sufficient to account for observable behaviour and the proximal causes – in particular the orderliness of human action.

- Methods in CSE must be applicable to the study of human-technology coagency, either for the purpose of understanding the nature of performance of the JCS or for the purpose of specifying and building JCSs with specific objectives and uses. In developing explanations, the methods describe how the performance of the JCS should be analysed; in building new systems, the methods describe the consequences of specific decisions and how this may affect the performance of the JCS and specifically the propagation of consequences. A specific method can prevent inconsistency when the classification scheme is used by different people, or applied by the same person working on different occasions.

To agree with the principles of the models, the classification scheme must enable clear and consistent descriptions of observable performance traits and the underlying regularities. For instance, it is more important to be able to describe the characteristics of remembering as a function than to describe the characteristics of memory as a structure. The former would focus on typical strategies and heuristics for remembering, while the latter would focus on the characteristics of various types of memory, such as a working store and a long-term store. As there is no complete vocabulary of CSE, it is impossible to present the classification scheme as a single, coherent arrangement. The terms and concepts that CSE uses will be obvious from the way in which the main themes of CSE are described throughout this book. Most of the terms will be easily recognisable and correspond to common usage. In some cases it has, however, been necessary to replace terms that have a customary interpretation within the information-processing paradigm with terms that are not normally associated with the study of human-machine ensembles. The purpose of this is to maintain the denotation of a term while avoiding undesirable connotations.

REQUIREMENTS TO MEASUREMENTS

One of the most important problems for the behavioural sciences – and for the empirical sciences in general – is the definition or specification of what should be measured. Performance measurements must ideally meet three requirements: (1) they must be possible; (2) they must be reliable; and (3) they must be meaningful and possibly also valid. Very few of the measurements that are used in practice meet all three requirements. All, of course, meet the first requirement – although suggestions sometimes are made for hypothetical measurements that are impossible either for philosophical or technological reasons (e.g., the cognoscope proposed by Crovitz, 1970). Many measurements meet the second requirement, and some the third as well.

One important distinction among measurements is whether they are theory driven or theory begging. A theory driven measurement is derived from an articulated model of a phenomenon or a functional relationship, in the sense that the semantics – or the meaning – of the measurement is provided by the model. A theory begging measurement is derived from an indistinct or incomplete model, often referred to as a folk model – using the term in a non-derogatory manner (Dekker & Hollnagel, 2004). A folk model expresses a commonly held idea or notion, often shared among experts and non-experts, about the nature of an everyday phenomenon. Folk models are common within psychology and the behavioural sciences, probably because everyone has 'privileged knowledge' about how the mind works (Morick, 1971). A folk model is not necessarily incorrect, but relative to an articulated model it is incomplete and focuses on descriptions rather than explanations.

Folk models are also very difficult to prove wrong; i.e., they are not falsifiable in Popper's (1959) meaning of the term. On the other hand, articulated models are not necessarily correct in the sense of being valid. The fact that a measurement can be defined and interpreted relative to a model does not mean that it describes something real. The history of (pseudo)science has many examples of that, from phlogiston theory to homeopathic potencies. The distinction between articulated models and folk models is nevertheless useful to evaluate some of the performance measurements that are commonly used.

Theory-driven Measurements

The definition of a measurement depends on how the domain is thought of, regardless of whether this is done implicitly or explicitly. A measurement is an expression, in quantitative or qualitative terms, of the value of a system parameter or characteristic – such as the level in a tank, the area of a surface, or the frequency of a specific event – at a given point in time or for a known duration. That which is measured must obviously be describable by the classification scheme and the definition of a measurement must be based on a model, which describes what the essential aspects of performance are and which therefore constrains what can be measured. The model parameters effectively become the basis for specifying the measurements and thereby help to define both how the measurements can be made and how they can be interpreted.

As discussed in Chapter 1, most operator models are structural in the sense that they explain behaviour by means of some hypothetical elements of mental machinery, as well as by the relations between them. A good illustration of that is the conventional information-processing model shown in Figure 3.1. Here human actions are described in terms of a relative simple set of functional units, which are called 'perception', 'cognition', and 'motor response' respectively. In the terms of this model, stimuli first enter the perception stage, where they are detected, recognized, identified, or categorised. Processing then moves on to the cognitive stage, where decisions are made and responses selected, possibly through a number of iterations, before finally passing on to the motor response stage. Individual stages can have memories associated with them, such as a short-term sensory store composed of visual and auditory image store, and a long-term memory with a working memory as a subset. The significance of this model is the implications for what can be measured and how measurements should be interpreted. Typical measurements have to do with the characteristics of the various stores and the transfer of information between them, the speed and capacity of the various processes, etc. It is also noteworthy that there are more details about the input side than about the output side.

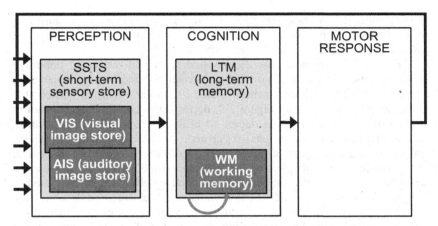

Figure 3.1: A characteristic information processing model.

Articulated models enable well-defined description of the functional relations of the target system, usually in the form of a theory or a set of hypotheses. Unfortunately, a well-defined description is not necessarily correct. Articulated models are typically structural and the measurements are related to the structural and functional characteristics of model elements. One reason for this is that it is more difficult to model actual work and actions than to describe a hypothetical mental mechanism. A mechanism, whether mental or technological, is stable and has easily identifiable components. Performance in a work situation – and even in a controlled experiment – varies over time, which makes it much more difficult to describe. One solution is to interpret performance diversity as variations of a procedural prototype, by invoking various convenient additional assumptions (cf., Hollnagel, 1993a). Another solution is to look for higher order regularities of performance, with only minimal assumptions about what may have caused them. CSE favours the latter approach, although it means that the measurement issue initially may be more difficult.

Theory-begging Measurements

If an articulated model is not available, measurements can be based on a general understanding of the characteristics of the system and of the conditions of human work. A classical example of that is fatigue (Muscio, 1921), workload (Stassen, Johannsen & Moray, 1990) and more recently situation awareness (Endsley, 1995). Workload, for instance, reflects the subjective experience of mental effort, which is so pervasive that it can be applied to practically every specific situation. Furthermore, workload is acknowledged to be an important causal factor in folk models of human performance. It is a measurement that is defined by consensus because it is easily understood by everybody. It is not defined by an articulated model; indeed, the models have usually come afterwards.

Folk models typically describe measurements that reflect an important aspect of the human condition but that refer to intervening variables – representing intermediate mental states – rather than to observable performance. As shown by Figure 3.2, performance can be characterised in many different ways, such as degree of efficiency, speed and precision of actions, performance deviations or 'errors', etc. Performance depends on individual attributes such as experience, motivation, capacity, skills, etc. Performance is also affected by working conditions, which in turn can be described by means of a number of general and specific factors. Folk models propose that one or more of the intervening variables are strongly correlated with the quality of performance, and that measurements of this intervening variable provide essential information about how well the individual is able to accomplish a given task. Some of the more common intervening variables are shown in Figure 3.2.

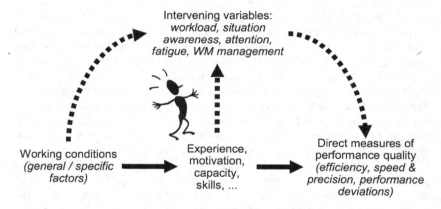

Figure 3.2: Loosely defined measurements.

In order for a folk model to be of practical use, it is necessary to assume that the measurement – or rather the hypothetical state that the measurement refers to – is affected by the working conditions in the same way that an individual being in the situation would be. (It also implies that the hypothetical state is the prime determiner of operator performance.) It must also be assumed that the measurement can be used to predict short-term changes in performance, since that is the main reason for not just observing what happens. The reasons for relying on the measurements are therefore (1) that it is easier to make them than to observe and analyse actual performance, and (2) that they are common to many different tasks and domains hence allow generalisations across problems. Both of these reasons point to obvious advantages in the study of human-machine ensembles. The main drawback of such measurements is that they have a vague theoretical basis and that their definition often lacks sufficient rigour. This problem has plagued behavioural science for the best part of a century, from Muscio's (1921) conclusion that a fatigue test was impossible to Moray & Inagaki's (2001) conclusion that complacency cannot be measured. In both cases an essential part of the main argument was that it is inappropriate to measure something unless the phenomenon in question (fatigue, complacency, etc.) can be established in some independent manner.

The Meaning of Measurements

The range of measurements that is typically employed in empirical research – particularly in controlled experiments – can be characterised in terms of two characteristics. The first is how easy it is to make the measurement, whether the data are readily available and can be reliably recorded with a reasonable effort. The second is how meaningful the measurement is, i.e., how easily it can be interpreted and how valid the interpretation is. It is an all too common experience in behavioural research that measurements are chosen because they are possible to make and because there is technology available, rather than because they are meaningful. Figure 3.3 illustrates the position for some of the more common measurement types on the two dimensions.

As Figure 3.3 shows, the various measurements seem to fit along a diagonal that represents a trade-off between ease of measurement and explicitness of the theoretical basis. Many measurements are relatively easy to make but have a limited theoretical basis and are therefore difficult to interpret. This is typically the case for measurements that can easily be recorded, such as physiological variables (e.g. heart rate) or overt performance (e.g. audio, video recordings). Keyboard interaction or keystrokes can be recorded with high reliability and little cost, but there is no articulated theory that makes keystroke data meaningful, although they are valuable as one source of raw data when combined with others. Other measurements have an acceptable theoretical foundation but are either difficult to make or difficult (and laborious) to interpret. Examples of that are eye movement recordings or performance 'errors'. Measurements such as workload and situation awareness occupy an intermediate position. It requires some effort and ingenuity to make the measurements, and, although it may be quite easy to transform them to meaningful statements, the interpretation is not supported by an articulated theory. It would clearly be very useful if measurements could be proposed that were both easy to make and theoretically meaningful. It follows from the preceding arguments that such measurements must be based on an articulated model, rather than a folk model. Unfortunately, there are few candidates for this category.

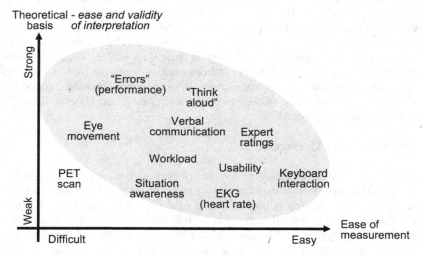

Figure 3.3: Dimensions of measurements.

THE ELUSIVENESS OF COGNITION

The study of human-machine systems has since the 1980s mostly been the study of human-machine interaction as practised by the human information processing school of thought. Cognition has either been treated as information processing or, at best, as an epiphenomenon of information processing. There are, however, different views on whether cognition should be studied under reasonably controlled conditions or under more natural conditions. Both views have a respectable pedigree in the history of psychology and anthropology/ethnology respectively. In the 1980s and 1990s the latter view became known as the study of 'cognition in the world' or 'cognition in the wild', which reasonably makes the former the study of 'cognition in the mind' or 'cognition in captivity'.

Cognition in the Mind

When psychology came into its own as a science towards the end of the 19th century, the introspective method favoured by philosophers also became the method preferred by Wilhelm Wundt and others. It was hoped that a meticulous reporting of inner experience would unveil the structure of the conscious mind. This approach was dealt a devastating blow by John Watson's behaviourist manifesto *Psychology as the Behaviorist Views It* (Watson, 1913). The consequence, particularly in the United States, was that researchers tried to avoid all topics related to mental processes, mental events and states of mind. Instead, animal psychology became fashionable, and a humongous amount of data concerning conditioning of behaviour was produced.

Although Behaviorism kept a firm grip on academic psychology for more than four decades, the ambition of studying cognition never completely vanished (Ritchie, 1953). As described in Chapter 1, developments in the meta-technical sciences seemed heaven-sent to psychologists looking for an alternative, since they offered a way in which cognition could be modelled as specific processes or functions without raising the spectre of mentalism. As early as 1946, Edwin Boring described a five-step programme for how the functions of the human mind could be accounted for in a seemingly objective manner. He argued that the ultimate explanation should be given in terms of psychophysiology, so that "an image is nothing other than a neural event, and object constancy is obviously just something that happens in the brain" (Boring, 1946, p. 182). This challenge was later taken up by cognitive science leading to the study of cognition as an inner – mental – process rather than as action, i.e., as 'cognition in the mind'. The idea of context-free cognition was promoted by leading scientists such as Herbert A. Simon, who argued very convincingly for the notion of a set of elementary information processes in the mind. One controversial consequence of this assumption was that the complexity of human behaviour was due to the complexity of the environment, rather than the complexity of human cognition (Simon, 1972). This made it legitimate to model human cognition independently of the context, which effectively was reduced to a set of inputs in the same way that actions were reduced to a set of outputs.

For reasons that have been described in Chapter 1, the study of human-machine interaction – and in particular the study of human-computer interaction – soon became dominated by the study of 'cognition in the mind'. This happened in good faith because it was commonly agreed that human information processing was not only necessary but also sufficient to understand human action. Although the value of this research in many ways is unquestionable, it gradually became clear that the study of 'cognition in the mind' was inadequate. One reason was a number of serious accidents that confronted researchers with the complexity that humans have to deal with in the real world. This was particularly striking since the main vehicle of research had been – and in many ways still is – the controlled experiment, which relies on a reduced representation of a real world problem (Hammond, 1993). In addition to the accidents, the growing number of studies also pointed to the obvious, but curiously overlooked, fact that people rarely work alone. On the contrary, people are always part of a group or an organisation, even though they may be separated from them in time and space. It was slowly realised that the study of cognition could not be pursued without including the socio-technical context. People act in a situation or a 'world' and their performance is determined as much by what they expect will happen as by what actually happens. Comprehensive studies demonstrated that people normally spend more effort – cognition – in preparing what they should do, than in actually doing it (Amalberti & Deblon, 1992). This can also be expressed by saying that it is more important to make sense of a situation than to make decisions in it. In hindsight this is obvious, since when a situation makes sense then there is no need to make decisions.

Cognition in the Wild

In cognitive science, the recognition of the need to go beyond what happened in the mind was awkwardly named situated cognition (e.g. Clancey, 1993). This maintained the focus on what happened in the brain – cognition as computation – but acknowledged that cognition emerged as a result of the interaction between the brain, the body and the world. This view logically abandons the idea a mental model as the exclusive basis for deciding on what to do. Instead, the person makes use of the information that exists in the situation and may even actively contribute to or enrich the external representation. In most cases there is actually no sharp line between what goes on inside the head and what happens in the world. People use whatever is available as an efficient scaffold or platform to reduce the demand for mental resources (Clark, 1997). Instead of studying subjects in laboratories under highly constrained conditions, situated cognition followed Brunswick (1956) by emphasising that data must be ecologically valid in the sense that they tell us something about how humans act in natural situations.

The commonly accepted term for the study of work in the actual technical setting is 'cognition in the wild' (Hutchins, 1995). A related term is naturalistic cognition and specifically naturalistic decision making (Klein et al., 1993). Typical examples of the settings that have been studied are flight decks of commercial airliners, control centres and control rooms of various kinds, surgical operating rooms, etc. Such settings require models that do not reduce the complexity to what can be expressed as input and output only, and methods that do not presuppose controlled experimental conditions. Despite the relatively tender age of this direction of research, the concepts and methods are quite mature, not least because they have benefited from the long experience in disciplines such as anthropology and ethnology, as well as from the steadily maintained European tradition of field studies in dynamic psychology and ergonomics.

The lesson learned from these studies is that work in real world conditions clashes with the assumptions of 'cognition in the mind'. In order to understand the performance of the human-machine ensemble, it is necessary not only to focus on 'cognition in the world' but to extend the meaning of cognition far beyond the notions of human information processing. Indeed, even the term 'cognition in the wild' is, in a sense, a misnomer since it implies that cognition is something that can be studied on its own. Semantically, the main difference between 'cognition in the mind' and 'cognition in the wild' relates to how cognition should be studied. From a CSE perspective, the question is whether one should study cognition *per se* at all, the alternative being to study the performance of the JCS (Hollnagel, 2001). In CSE, human cognition is no longer the central issue, regardless of whether it is assumed to exist in the mind or in the world. Since JCSs can range from human-technology ensembles to organisations, it is somewhat presumptuous to assume that the anthropomorphic perspective provides the right place to start. There may, indeed, be commonalties among JCSs at different levels, but they are more likely to be at the level of performance than at the level of the underlying processes. In practice that probably also agrees with the intention of studying 'cognition in the wild', and the continued use of the term cognition is more than anything else due to terminological hysteresis.

THE FOCUS OF CSE

Since CSE is focused on performance, the main interest is in finding out what the regularities of performance are, and how such regularities – as well as possible irregularities – can best be described and explained. In contrast to cognitive engineering and the information processing approaches, it is not assumed *a priori* that cognition, whether human or artificial, is the most important determinant. For example, the behaviour of a human problem solver is interesting because it contributes to the performance characteristics of the JCS, but not because problem solving by definition is an interesting process. The bulk of the evidence from studying 'cognition in the wild', indeed, lends support to the view that the influence of working conditions and the situation as a whole are greater than the influence of human cognition seen as the processes of thinking and reasoning. The unit of analysis must therefore be the performance characteristics of the JCS rather than the cognition that may – or may not – go on inside. Concretely, the focus of CSE is on how the JCS is able to maintain control of a situation, specifically how it is able to accomplish its functions and achieve its goals. The focus is consequently on the subject-*cum*-context conglomerate, rather than on the subject and context *dyad*.

The analysis of a JCS must start by analysing the functions and activities that take place – or may take place if the system is not yet in operation. A function analysis can be based on a distinction between goals and means, which can be found in many scientific disciplines. It is also a fundamental conceptual device for system analysis, whether for JCSs or purely technological systems. The goals-means analysis has a long history, going back at least to Aristotle, who in Book III of the *Nicomachean Ethics* wrote:

> We deliberate not about ends but about means. ... (People) assume the end and consider how and by what means it is to be attained; and if it seems to be produced by several means they consider by which it is most easily and best produced, while if it is achieved by one only they consider how it will be achieved by this and by what means this will be achieved, till they come to the first cause, which in the order of discovery is last. ... (And) what is last in the order of analysis seems to be first in the order of becoming. ... The subject of investigation is sometimes the instruments, sometimes the use of them; and similarly in the other cases – sometimes the means, sometimes the mode of using it or the means of bringing it about. (Aristotle, 350 B.C.)

We shall not at this point go into the details of the goals-means decomposition as a method, since this will be taken up in Chapter 6. Suffice it to say that a functional analysis, a goals-means decomposition, provides a powerful tool that does not require any prior assumptions about the nature of cognition or about what the primitive processes are.

Cognition and Context

One of the most popular terms in the 1980s and 1990s in cognitive psychology was *context*, with the corresponding notion of situated cognition. Considering cognition and context together is not only appropriate, it is imperative and therefore in a sense superfluous. To talk about cognition and context – such as contextual or situated cognition – logically implies that there is also cognition without a context, i.e., cognition as a pure, abstract, or ethereal process. Since this clearly cannot be the case in an absolute sense, there is strictly speaking no need to talk about context.

The reason for the confusion has something to do with the traditional study of cognition, which, following the practice of experimental psychology and empirical science, was based on controlled laboratory environments. Just as Galvani studied the functions of a frog's muscle by itself – without the rest of the frog – and just as Wundt, Ebbinghaus and a host of other psychologists tried to study psychological functions in isolation, so has cognition been studied in splendid isolation (Hammond, 1993). This tradition was reinforced by Behaviorism (the American spelling is used deliberately) and continued by the early cognitive psychology (e.g. Neisser, 1967). The tradition was almost broken in the late 1970s (Neisser, 1976), except for the sweeping success of artificial intelligence (AI). Most people seemed oblivious to the fact that AI was successful only when it studied isolated cognitive or intellectual functions in simplified environments, such as the blocks' world (Winograd, 1972) or STRIPS (Fikes & Nilsson, 1971). Human cognition was seen as a set of information processes that took place inside the head of a person, and which could be studied in their own way, i.e., by carefully observing how a person performed in well-defined situations.

The mistake of the dominant tradition in cognitive psychology and cognitive science has been to confuse the fact that there are general characteristics of cognition (common traits, features across situations) with the assumption that cognition is an isolated, context-free process. Instead, the general characteristics simply reflect the relative constancy of the environment. (This, however, does not commit us to accept Simon's conclusion that human cognition is simple.) The notion of a context-free environment or condition rather means that it is highly simplified and generalised, i.e., that it is valid for many different specific situations. Conversely, something that is context bound is valid only for one set of conditions. But even if this definition is adopted, it must be realised that nothing is ever context-free. Even in highly simplified situations there is a context, i.e., a set of conditions and assumptions – spoken or unspoken. These conditions and assumptions form the basis for the actions and expectations of people.

Cognition and Control

Two things characterise human beings: one is the capacity to learn; the other is the capacity to adapt (and to plan). Both are essential characteristics of cognitive systems and together they mean that there necessarily must be a context. Learning is inconceivable without a context, i.e., a world or an environment in which actions unfold. Continued learning is, furthermore, necessary because the context is dynamic and is changing. If the conditions were fixed, learning would be necessary only up to a certain level, sufficient to ensure survival. The same line of reasoning can be applied to argue the necessity of adaptation.

The capacity to learn is tremendous and pervasive, and so is the ability to adapt. But both learning and adaptation require efforts, and natural cognitive systems cleverly try to reduce the need of both. Adaptation is needed mostly when the situation is new. Conversely, if a situation is recognised, adaptation proper is no longer required. Instead, we can rely on already existing plans or skills to carry us through – although possibly with some modifications or local adaptations.

To illustrate this, consider the difference between a neonate and an adult. We can assume that nothing can be recognised by the neonate, hence that nothing can be based on learned patterns. Furthermore, the behavioural repertoire of a neonate is very limited, because control of limbs and motor apparatus is rudimentary. A neonate can therefore survive only if the environment provides conditions that are suitable for the fragile organism and shows little variation, something that parents and societies are very good at. However, as soon as something is learned, situations or contexts become familiar and reactions can be based on learned patterns, thereby tolerating gradually larger variations. In the adult the reliance on patterns is often taken to the extreme. Part of the educational and vocational training system is, indeed, to instil patterns and responses in people so they can take appropriate action without having to spend time and effort to adapt. Adults tend first to look for familiar patterns that can be used as a basis for reactions, and resort to adaptation only when absolutely necessary.

This approach is highly efficient as long as the environment is relatively stable and predictable. In cases where the environment or the situation is unfamiliar, relying on pattern recognition may lead us astray. One consequence – or even a definition – of being in an unfamiliar situation or environment is that unexpected events may happen, hence that relying on pattern recognition is insufficient. Unexpected events usually mean that the current understanding is incomplete and that it therefore must be revised. An incorrect understanding may easily lead to inappropriate actions being chosen; these may in turn lead to more unexpected events or unexpected results, which challenge the current understanding, and so on.

For CSE, cognition is the process by which we cope with the complexity of the situation. As we mature, coping is based on recognition of situations and stereotyped responses – possibly modified at the local level by small-scale adjustments. If the local adjustments fail then coping is based on adaptation, but even adaptation makes use of generalised response patterns and is in a sense bootstrapped from them. It is only if everything fails that we must resort to the basic adaptation strategy known as trial-and-error type of performance. For CSE these aspects or regularities of performance are important to study, as they occur in cognitive systems at various levels – natural or artificial, simple or complex.

Modelling Cognition and Context

In order to be able to understand how a JCS functions, models should be at the level of meaningful system behaviour rather at the level of the underlying processes. Modelling therefore cannot be of cognition alone but must be of cognition and context as a whole, or in other words of coagency.

Chapter 1 characterised the information processing models as procedural prototype models (Hollnagel, 1993b). These models assume the existence of an internal representation of a characteristic sequence of actions as a basis for the observed sequences of actions. The classical example is the sequence of processes that go on between input (S) and output (R) – or in other words the details of the functioning of the organism (O), for instance the information-processing model shown in Figure 3.1. In such models, changes in the environment are described as changes in the input to the internal processes, but these changes do not affect how the steps are ordered; their assumed 'natural ordering' is maintained from one event to the next. Although bypasses or variations/deviations from the prototype sequence may skip one or more steps, the underlying ordering is immutable, as the following description makes clear:

> Rational, causal reasoning connects the 'states of knowledge' in the basic sequence. ...Together with the work environment, the decision sequence forms a closed loop. Actions change the state of the environment, which is monitored by the decision maker in the next pass through the decision ladder. (Rasmussen, 1986, p. 7)

In this description, the decision ladder is the procedural prototype. Yet years before the adoption of the IPS as the pervasive metaphor, a leading behavioural scientist, Karl Lashley, summarised his research as follows:

> My principal thesis today will be that the input is never into a quiescent or static system, but always into a system which is already actively excited and organized. In the intact organism, behavior is the result of interaction of this background of excitation with input from any designated stimulus. (Lashley, 1950, p. 458)

Indeed, Lashley concluded that it was possible to understand the effects of a given input only if the general characteristics of this background of excitation could be described. Lashley's research was not with work but with the nature of memory traces, yet even here there was a recognized need to invoke the context to understand what happened.

In order properly to model cognition and context it is necessary to abandon the IPS metaphor and the notion of a procedural prototype. The alternative, called contextual control models, focuses on what a JCS does, specifically how actions are chosen to comply with the current goals and constraints. Contextual control models are based on concepts that are applicable to all types of cognitive systems, even though some of the terms have an undeniable psychological flavour. (On the other hand, it is not unusual to use anthropomorphic terms to describe machines.) Contextual control models focus on macro-cognition rather than micro-cognition (e.g. Cacciabue & Hollnagel, 1992; Klein et al., 2003) and are by their nature cybernetic rather than cognitive. The context exists both as the work environment and as a construct maintained by the JCS. The term *construct* emphasises that this representation is of a temporary nature and that it has been deliberately put together to be the basis of purposive action. The construct is essential to determine what the next action will be and therefore at the same time reflects the context and creates it.

Sequentiality in Reality and in Models

Procedural prototype models imply that behaviour is the result of an orderly process that can be represented as a series of stages. This bias comes about because the fact that the description of any set of events – past or future – necessarily must be ordered in time is misinterpreted to mean that the events therefore also arise from a sequential process. Since this bias often goes unrecognised, it is worthwhile spending a little more effort to discuss it, particularly as it pertains to the modelling of cognition.

Ever since Behaviorism was renounced by mainstream academic psychology, it has been a truism that human behaviour could not be explained without understanding what went on in the human mind. Models of what goes on in the mind – whether it is called cognition or information processing – are primarily based on two types of raw data, introspection and think-aloud protocols, where the latter in turn rely on the introspection of the subject. Introspection invariably describes events that happen in a sequence, but this sequentiality is due to how the description comes about rather than to what is described, i.e., because introspection can describe the conscious experience of only one thing at a time.

In addition to the impossibility of reporting simultaneous conscious experiences, it is also a fundamental fact that events do happen one after the other. This is simply a characteristic of the physical universe by which we exist, in which time is one-dimensional. (We are, of course, talking only about macroscopic events, since the rules of the quantum world are quite different.) For events that have happened, the ordering in time is given on either an interval or ordinal scale and one of the central steps in experiment analysis is indeed to order the data with respect to a time-line (e.g., Sanderson, Watanabe & James, 1991). For events that will happen in the future, we may not always know their order or be able to predict it. (Note, however, that for the design of artefacts it is usually essential to ensure that things happen in a pre-defined order.) Yet because of the nature of our universe it is certain that they will happen in sequence, rather than all at the same time. The sequentiality bias also affects the representation of what has been reported. All written representations, such as verbal protocols, are by their nature sequential. Graphical representations may be able partly to overcome this, but even though sub-processes may be shown in parallel, the sub-processes themselves are sequential.

Altogether these arguments make it clear why the representation of what goes on in the mind must include an indication of sequence or temporal order. Yet this does not warrant the conclusion – or inference – that there is an underlying reason for it, in the sense that the selection of actions or events is inherently sequential. As a parallel, consider the outcomes of a random process such as a series of throws of a die. The description of the outcomes is invariably given in a sequence, since the outcomes occurred one after the other. Yet this does not mean that the underlying process was sequential, since each throw of the die was independent of the previous. The assumption that behaviour is the result of an orderly process that can be represented as a series of stages is due to an understandable bias, but a bias nonetheless, and one that should be avoided if at all possible.

THE THREADS OF CSE

CSE is concerned with how human-machine coagency can best be described and understood. The focus of CSE is defined by the driving forces and by the nature of work, as described in the preceding chapters. In consequence of that, CSE is focused on the following threads or themes:

- How people *cope with* the *complexity* that is a result of the many technological and socio-technical developments. This issue has become particularly potent in light of the pace of development in the last half of the 20th century, which shows no indication of slowing down.
- How people *make use of artefacts* in their work (and at leisure), specifically how the use of artefacts has become an intrinsic part of intentional activity. The slightly cynical versions of that would be to ponder how artefacts make use of people, or perhaps how the technological development can misuse or mistreat people.
- How humans and artefacts can be described as *joint cognitive systems*, and how this extends the scope from the focus on the interaction between human and machine to how humans and technology effectively can work together.

The three main threads have all been mentioned in the preceding. Each thread will be described separately in the three following chapters.

Coping With Complexity

The operators' work in the control of complex, dynamic processes has been described as coping with complexity (Rasmussen & Lind, 1981). The complexity in process control is due to the multiple channels of information and lines of control through which work has to be done, as well as to the possibly conflicting goals that characterise the operators' working situation. Today the very same characterisation may fit a much larger number of jobs, hence making coping with complexity more pervasive. The solutions that operators apply typically aim to reduce complexity, for instance by structuring the information at a higher level of abstraction and determining the appropriate actions at that level. The development of information technology has at the same time increased the complexity for many types of jobs, and made it possible to computerise the analysis of incoming information, thereby, to some extent, accommodating the operators' needs, hence helping them in their coping. This kind of support clearly involves a replication of parts of human cognition in the machine and therefore represents the use of an artificial cognitive system.

As described by the cyclical CSE model, a JCS tries at all times to choose the actions that will achieve its goals. The choices depend on the resources, capabilities, and competence of the JCS, which entails the operator and the supporting technology, as well as on the construct, i.e., the current situation comprehension. The choices also reflect the restrictions or constraints that exist in the situation, such as the available time and general resources, prevailing conditions (and possible disrupting influences), etc. The complexity comes about because neither goals, nor resources, nor constraints remain constant. In fact, the constraints in particular may change as a result of the actions that are chosen, hence introduce a dynamic coupling between the JCS and its context.

Use of Artefacts

Work, with only a few exceptions, involves the use of artefacts (or natural objects) with the purpose of achieving a specific goal. An artefact is a device used to carry out or facilitate a specific function that is part of work. The history of artefacts has in many ways been a history of amplification, as recounted in Chapter 2. Although some artefacts in support of cognition have existed for ages, such as memory aids, the use of computers and information technology has made it possible to design artefacts for more sophisticated functions such as collating, forecasting, decision making and planning. Some artefacts may even be rudimentary cognitive systems themselves, in terms of being able to maintain control. The use of artefacts to support cognitive activities has changed from the metaphorical to the real, and this creates a need for proper concepts and methods.

One of the goals of CSE is to understand how people individually, as a group, and as operational organisations shape artefacts to serve their purpose. Research in this area must therefore consider the differential impact of artefacts on the practitioner's work and coping strategies (Woods, 1998). Since representations of the problem domain available to the practitioner can either undermine or support strategies related to task performance, there has for some time been a pressing need to develop "...a theoretical base for creating meaningful artefacts and for understanding their use and effects" (Winograd & Flores, 1986, p. 10). In order to do this CSE must confront the difficult problem of the interaction and interdependence of investigations and design. Effective design depends on the results of investigations of the cognitive activities in the field of practice, but it is as yet unclear how these can be utilised when the artefacts may transform the JCS in question in potentially radical ways.

Joint Cognitive Systems

Although classical ergonomics emphasised the necessity of viewing humans and machines as parts of a larger system, the distinction between the operator as an intelligent human being and the machine or the process as a technological system was never questioned. As we have seen, this seemingly innocuous and 'natural' decomposition meant that the interaction between human and machine assumed prominence and that the overall system perspective consequently was neglected. CSE overcomes that by shifting the focus from human and machine as two separate units to the JCS as a single unit. It follows from this that what should be studied is neither the internal functions of either human or machine nor the interaction between them, but rather the external functions of the JCS as based on human-machine coagency. This change is consistent with the principle that humans and machines are 'equal' partners, and that they therefore should be described on equal terms. Humans should not be described as if they were machines, nor should machines be described as if they were humans.

An important consequence of focusing on the JCS is that the boundaries must be made explicit, both between the system and its environment and between the elements of the system. A system is generally defined as "a set of objects together with relationships between the objects and between their attributes" (Hall & Fagen, 1969, p. 81) – or even as anything that consists of parts connected together. In this definition, the nature of the whole is arbitrary, and the definition of the system is therefore also arbitrary.

In the case of a JCS, the boundary clearly depends both on the purpose of the analysis and on the purpose of the JCS. In the view of CSE, a pair of scissors would not be called a cognitive system, although it certainly is an artefact that under some conditions can be used to augment human cognition (cf. Chapter 5). A woman using the scissors is obviously a cognitive system and the woman-*cum*-scissors is therefore a JCS. The advantage of using a pair of scissors as an example of an artefact rather than, e.g., a computing system, is that it is easier to see that the woman is *using* the pair of scissors rather than *interacting* with it. It is meaningful to consider the woman-*cum*-scissors as a system as such, but also to consider this system as part of a larger system, for instance, the manufacturing cell. That may in turn be described both as a JCS in itself, and as a part of an even larger system, and so on. The point of Beer's (1964) definition, and of system definitions in general, is to emphasise that there are no natural systems in an absolute sense – although there well may be habitual delimitations of some systems. Something is called a system if it is seen relative to an environment, and separated from this by a boundary.

In Chapter 1, a cognitive system was defined by its ability to modify its behaviour on the basis of past experience so as to achieve specific anti-entropic ends. A JCS obviously has the same ability. The difference between the two is to some degree a matter of distinction, which really becomes obvious only on the level of the single individual: a human being is a cognitive system, but not a JCS; a human using an artefact is a JCS; a group of people, even as small as two such as the pilot and co-pilot in the cockpit, is a JCS. The implication of calling it joint is that it can meaningfully be decomposed into parts, of which at least one is either a cognitive system or a JCS. Other parts may be artefacts that are not cognitive in terms of the above definition. An organisation or a group of people can be decomposed in this manner, sometimes even through several steps. But a single human being cannot be decomposed in this way, since it does not make sense within the domain of CSE to consider human subfunctions (e.g. temperature regulation or contour recognition) as cognitive systems *per se*. More formally, a JCS can be defined as in Table 3.1, using the Backus-Nauer Form (BNF) syntax.

Table 3.1: A BNF Definition of a Joint Cognitive System

JointCognitiveSystem ::=	<CognitiveSystem> <JointCognitiveSystem> | <CognitiveSystem> <CognitiveSystem> | <CognitiveSystem> <Artefact>
<Artefact> ::=	<PhysicalArtefact> | <SocialArtefact>
CognitiveSystem ::=	Simple system capable of anti-entropic behaviour
PhysicalArtefact ::=	Object produced or shaped by joint cognitive system
SocialArtefact ::=	Rules or regulations produced by joint cognitive system

Merging the Threads

As the above summaries hopefully have made clear, the three threads of CSE are not independent. Although for practical reasons it may be necessary to focus on one at a time, this should not lead to the others being forgotten. Indeed, the three threads indicate three different ways of looking at the same problem, rather than three different problems. This can be illustrated by relating them to the basic cyclical model, as shown in Figure 3.4.

In terms of Figure 3.4, coping with complexity can be seen as the ways in which people try to make sense out of the information that is available to them – and often forced upon them. In Figure 3.4 that is shown as the relation between event/feedback and current interpretation, but it also covers the complexity of choosing what to do, specifically when the action involves the use of a non-trivial controls. That leads to the next thread, the use of artefacts. This refers primarily to the use of artefacts in the implementation of actions, hence the relation between current interpretation and choosing the next action as shown in Figure 3.4. Yet as argued above, artefacts are not just passively used, but do also themselves have a determining effect on human performance and how the world is seen. The consequence of that is that we need to go beyond describing how humans interact with artefacts, and instead consider how the human-artefact ensemble performs, i.e., the issue of JCSs. In Figure 3.4 this is shown as covering all the aspects of the cyclical mode, and as providing a way of relating coping with complexity to the use of artefacts.

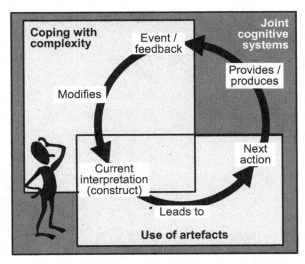

Figure 3.4: Merging the threads of CSE.

Chapter 4

Coping with Complexity

The basic issue for CSE is how to maintain control of a process or an environment. Both processes and environments are dynamic and therefore complex, and joint cognitive systems are striving to cope with this complexity. The coping takes place at both the individual and organisational levels, the latter as in the design of work environments, of social structures, and of technological artefacts.

INTRODUCTION

Coping with complexity is about the effective management or handling of situations where there is little slack, where predictability is low, and where demands and resources change continually. Coping with complexity is thus essential to maintain control. Since it is clearly desirable that a JCS can do this for a reasonable range of performance conditions, design must consider both the variability of system states and the variability of human performance whether as individuals or teams. The purpose of design can, indeed, be expressed as enhancing the capability to be in control, thereby ensuring that the social and technological artefacts function as required and as intended.

About Coping

An early reference to the concept of coping with complexity can be found in the field of management cybernetics (Beer, 1964). Yet coping with complexity did not receive much attention until it was brought into the human-machine research vocabulary. Here it was defined as the ability "to structure the information at a higher level representation of the states of the system; to make a choice of intention at that level; and then to plan the sequence of detailed acts which will suit the higher level intention" (Rasmussen & Lind, 1981, p. 9).

Almost half a century ago Donald MacKay (1968; org. 1956) defined the minimal requirements to a system that would be capable of goal-guided – or goal-directed – activity. As shown in Figure 4.1 only three functional elements are needed, which MacKay called a receptor, a comparator, and an effector. (Notice that this is an elementary feedback-controlled system.)

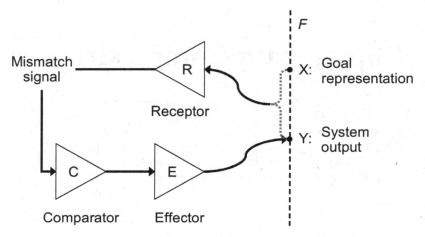

Figure 4.1: Minimal requirements for goal-guided activity (after MacKay, 1968/1956).

In Figure 4.1, the point X represents the system's goal and the line F the space of possible goal states. The active agent is represented by the effector system E, which is governed by the control system C. The system is said to be goal-guided if the overall pattern of E's activity reduces the distance between Y and X to a minimum, within a given time interval. The distance between Y and X is expressed as a mismatch signal from the receptor system R.

> We assume that in a typical situation E is capable of a certain range or variety of modes of activity (including inactivity), so that the function of C is to select from moment to moment what E shall do next, out of the range of possibilities open to it. (MacKay, 1968, p. 32)

This simple system works well as long as the 'movements' of the target X can be represented as changes in position along a single dimension and as long as they are smooth or continuous. As soon as the 'movements' become more irregular, corresponding to a more complex and therefore also less predictable environment, more complicated functions are required to maintain control. MacKay argued that although having a hierarchy of controllers in principle could extend the same fundamental arrangement to cover the more complex situation, it became impractical to describe the performance of the system in terms of nested feedback-loops as soon as there were more than a few levels.

SOURCES OF COMPLEXITY

Attempts to define complexity are many and range from the useful to the useless. Complexity is never easy to define, and the term is therefore often used without definition. A start is, of course, to consider the dictionary definitions, according to which something is complex if it consists of (many) interconnected or related parts or if it has a complicated structure (sic!). A more substantive treatment can be found in the field of general systems theory, where complexity is defined by referring to the more fundamental concept of information. It is here argued that all scientific statements have two components. One is an *a priori* or structural aspect, which is associated with the number of independent parameters to which the statement refers. The other is an *a posteriori* or metrical aspect, which is a numerical quantity measuring the amount of credibility to be associated with each aspect of the statement. Complexity is now defined as follows:

> The amount of this 'structural' information represents what is usually meant by the complexity of a statement about a system; it might alternatively be defined as the number of parameters needed to define it fully in space and time. (Pringle, 1951, p. 175)

Pringle goes on to point out that the representation of complexity in the above sense is *epistemological* rather than *ontological* because it refers to the complexity of the description, i.e., of the statements made about the system, rather than to the system itself. Ontological complexity, he asserts, has no scientifically discoverable meaning as it is not possible to refer to the complexity of a system independently of how it is viewed or described.

This important philosophical distinction is usually either taken for granted or disregarded. In the latter case the epistemological and ontological aspects of complexity descriptions are mixed, which sooner or later creates problems for descriptions. The reason is that while the epistemological aspects are amenable to decomposition and recursive interpretation, the ontological aspects are not. Indeed, if complexity as an ontological quality of a system could be decomposed, it would in a sense be dissolved, hence cease to exist.

Some of the important factors that affect complexity are shown in Figure 4.2, superimposed on the basic cyclical model. This also suggests a convenient way to group the different factors.

- In relation to the evaluation and interpretation of events, two important factors are insufficient training and lack of experience. Of these, insufficient training is the more specific and also the one that best can be controlled by an organization. Shortcomings in the evaluation and interpretation of events may lead to an incomplete or partial understanding of the situation.

- Other factors are insufficient time and insufficient knowledge. Even if a condition can be recognized, it may be impossible to maintain a correct understanding, if time or knowledge are in short supply. This is particularly important for situations that are out of the normal, such as accidents. An incomplete or partial understanding leads to problems in choosing or selecting actions.

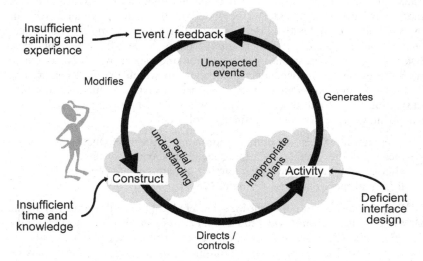

Figure 4.2: Important factors that affect complexity.

- A third group of factors is associated with the complexity of the interface, which both provides the information about what happens and the means by which an intended action can be carried out. If the interface is difficult to use, the implementation of an action may be incomplete or incorrect, leading to unexpected results. This problem is often solved by relying on a standard that is effective across different levels of user experience and cultures.

As already mentioned, the complexity of a process, hence the difficulties in coping, depends in the main on two closely coupled issues. One is the degree of orderliness or predictability of the process, and the other is the time that is available. The coupling between the two comes about in the following way. If predictability is low, then more time is needed to make sense of what is going on and to decide on the proper control actions. Conversely, if time is short or inadequate, it may not be possible to develop an adequate understanding of what is going on (develop an adequate construct), and control actions are therefore more likely to fail in bringing about the desired change. This will increase rather than decrease the unpredictability of the process, hence limit the available time even further. This particular kind of coupled dependency is technically known as a deviation-amplifying loop (Maruyama, 1963).

LOSING CONTROL

If we consider joint cognitive systems in general, ranging from single individuals interacting with simple machines such as a driver in a car, to groups engaged in complex collaborative undertakings such as a team of doctors and nurses in the operating room, it soon becomes evident that a number of common conditions characterise how well they perform, and when and how they lose control, regardless of domains. These conditions are lack of time, lack of knowledge, lack of competence, and lack of resources (cf. Figure 4.3).

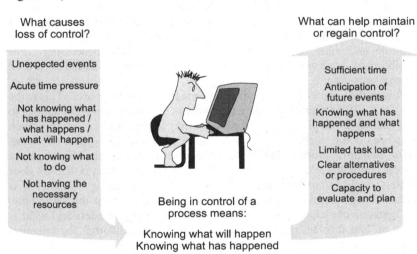

Figure 4.3: Determinants of control.

Of greatest importance is the intimate link between loss of control and the occurrence of unexpected events, to the extent that one might say that this in practice is a signature of loss of control. Unexpected events also play a role in another related way, namely as a consequence of lost control. The loss of control is nevertheless not a necessary condition for unexpected events to occur. There may be other factors, causes and developments outside the boundaries of the JCS that lead to events, which for the JCS are unexpected. These issues will be discussed again in Chapter 7.

Just as there are a number of generic conditions that make coping difficult and thereby may precipitate the loss of control, there are also some conditions that can enhance or facilitate coping and thereby prevent the loss of control – or perhaps even help in regaining control when it has been lost. As indicated by Figure 4.3, these are generally the converse of the conditions that lead to a loss of control. They will be described in more detail towards the end of the chapter.

Lack of Time

The occurrence of unexpected events is closely related to the amount of time that is available to the JCS in two different, but complementary ways. Firstly, the accuracy of predictions about what is likely to happen depends, among other things, on the available time. If the predictions are inaccurate, then unexpected events are more likely to follow. Secondly, unexpected events typically require additional time, since they fall outside the planned or customary activities. The operator needs time to take in the new information, to decide what to do about it, and to update the current understanding. This dilemma could in principle be solved if the JCS had infinite capacity, e.g., by invoking additional resources (corresponding to parallel processing). The possibility of increasing capacity on demand is, however, bounded even for a large organisation.

It is in the nature of *any* activity that time is a limited resource. This is the case for a JCS in control of a dynamic process or environment and applies to such examples as flying an aircraft, driving a car, trading on the stock exchange, doing surgery, or even cooking spaghetti Bolognese at home. Every dynamic process has certain demand characteristics because the process continues to develop and possibly changes state, even if the JCS does nothing. Unlike non-dynamic systems, such as a library search, it is impossible to suspend action and maintain the *status quo* at the same time. The concrete consequence is that there is limited time to understand what is going on and to plan, prepare, and execute actions. Although the temporal demand characteristics may be less pressing in some cases than in others, time is also limited in cases where the dynamic nature of the environment are less obvious, such as writing a scientific paper or planning the vacation next summer. The discipline of CSE is, of course, most interested in situations where time demands are tangible and where there are consequences at stake for the individuals, groups and organisations that are involved (cf. Chapter 3). This means cases where the rate of change of the process or target system is large relative to the duration of the activities under consideration. If unexpected events occur occasionally, there may be time and resources to cope with them without disrupting the ongoing activities and adversely affecting the ability to maintain control. But if unexpected events are numerous and if they cannot be ignored, they will interfere with the ongoing activities, hence potentially result in a loss of control.

Lack of Knowledge

In addition to having sufficient time, the JCS must also have sufficient knowledge of what is going on. It must specifically be able to recognise or identify what happens as well as to interpret it in a context. In the more formal terms of the Model-Classification-Method framework discussed in Chapter 3, the JCS must have a classification scheme or a set of categories by which the observed events can be described, which in turn requires that there must be an underlying conceptual basis or model.

With insufficient or no knowledge, all the time in the world will not make any difference. To take an extreme example, if you put a person 'from the street' into a control room of a refinery, he or she will have no real chance of controlling the process. To take a less extreme example, if you put people in charge of a job without giving them proper training and instructions, then they are likely to lose control of the situation and do things that lead to accidents, as in the accident at the JCO facility in Tokai-mura in September 1999 (Furuta et al., 2000). Control may also be lost in situations where people do not know what happens, since not knowing what happens affects the ability to predict. This can, of course, be seen as a case of unexpected events, yet in this case the events are not unexpected because they are deviations or interruptions, but because the operator does not know what is going on in the first place. Automation surprises (Woods & Sarter, 2000) are a typical example of that.

Lack of Readiness or Preparedness (Competence)

Even if developments are predictable so that few unexpected events occur, even if the operator knows what happens so that events and changes are easily and clearly recognised, and even if there is sufficient time, control may still be lost if the operator does not know what to do. Knowing what happens means that the recognition of events is fairly fast and efficient. Yet not knowing what to do means that the selection of actions may be difficult and take more time than usual – thereby eating into the available time. In the extreme case it is the well-known situation of realising the inevitable without being able to do anything about it. This need not be due to a lack of capability of the people in the system but of the system itself, in the sense that it has gone beyond the design limits. Well-known and tragic examples of that are TMI (where the lack of competence was reinforced by a misdiagnosis), Chernobyl, and the Concorde crash of July 25, 2000. Not knowing what to do is, of course, compounded by not having the time to do it.

Lack of Resources

Finally, control may be lost if necessary resources – other than time – are lacking. A lack of resources can be the result of an acute condition, such as a loss of power or pressure, or the consequence of a systemic failure, for instance the result of latent conditions or a mistake at the blunt end. One example is the recent the spread of the influenza in Sweden in the winter of 2001. In early January it was realised the influenza epidemic was coming but that all the vaccine had already been used, hence that a number of people would be left exposed. An unfortunately more frequent example is wildfires. Here the fire can be regarded as a dynamic process, which is reasonably predictable (although random events such as changing winds may play an important role). In the case of a wildfire there are few problems in knowing what is going on or what should be done, but the limiting factor is resources needed to extinguish the fire. The lack of resources means that a wildfire may go out of control, like the bushfire near Sydney in December, 2002.

COPING STRATEGIES

Coping with complexity can be expressed more formally by invoking feedback and feedforward control. If a process is controlled by means of feedback, the operator must have the capacity (and time) to make use of the information provided (as signals, measurements, or status functions). If a process is controlled by feedforward, predictions must be accurate and be ready when needed. Control that relies exclusively on feedback is likely to fail when time becomes too short, simply because the feedback cannot be processed fast enough to be of use. Control that relies exclusively on feedforward is likely to fail unless the predictions always are accurate, which is often impossible to achieve even for moderately complex environments. In practice, effective control therefore requires a balance between feedback and feedforward, something that can also be seen from the coping strategies that have been proposed over the years.

One example of that can be found in the studies carried out by Dörner and his associates (Dörner, 1980), which led to the formulation of a number of characteristic strategies. When dealing with complex processes (although in this case there were micro-world simulations), people tend to neglect how processes develop in time (what the state will be in the future). They also have great difficulty in dealing with exponential developments, and are inclined to think in causal series instead of causal nets, which means that they focus on the aspired main effect and neglect to consider the possible side effects. Examples of specific strategies are thematic vagabonding (jumping from one topic to the next, treating all superficially), encystment (sticking to a subject matter, enclosing themselves in it, treating small details very fondly, and not taking on anything else), and a decreasing willingness to make decisions or postponing them to a later time (Dörner, 1980).

A more inclusive view of this issue is that cognition may oscillate between intuition and analysis, where the latter requires more effort than the former. According to Hammond (1993), this was first proposed by the American philosopher Stephen Pepper in his book on *World Hypotheses* (1942). The juxtaposition of the intuitive and the analytic can also be found in decision theory, where satisficing (Simon, 1955) and 'muddling through' (Lindblom, 1959) represent the intuitive side while rational or economic decision making (e.g., Edwards, 1954) represent the analytical.

Information Input Overload

One aspect of coping with complexity is to be able to make use of the available information in time, i.e., fast enough to allow an action to be taken before it is too late. This ability is reduced in conditions where there is more information than can be handled, known as information input overload (Miller, 1960). Information overload is clearly a relative rather than an absolute condition, i.e., it means that the current capacity for dealing with or processing the information is insufficient. Overload thus occurs if the rate of input is increased, if the processing capacity is decreased, or if both happen at the same time.

Despite the obvious significance of this condition, there are only few direct studies of it, such as the one by Miller (1960) – which gave rise to the name – and Hiltz & Turoff (1985). The phenomenon is perhaps so widespread that it is taken for granted. The common reactions to information input overload range from the temporary non-processing of input to abandoning the task completely, similar to what later was called the exculpation tendency (Dörner, 1980). In between these extremes are a number of different strategies, which on the whole aim to preserve the essential information without causing the JCS to lag so much behind that the situation deteriorates (Table 4.1). The reactions can thus be seen as expressing a trade-off between demand and resources, which for humans mainly mean available time. What is essential information cannot be an objective category, but depends on the person's current interpretation of the situation (the construct) and the current goals. The very existence of these reactions implies a rather tight coupling between the information that is potentially available, the information that the person looks for, and the way in which this information is treated, cf. the cyclical model.

Table 4.1: Coping Strategies for Information Input Overload

IIO strategy	Definition	Criterion for use
Omission	Temporary, arbitrary non-processing of information, some input is lost.	If it is important to complete the task without further disturbances.
Reduced precision	Trading precision for speed and time, all input is considered but only superficially; reasoning is shallower.	If it is important to reduce or compress time, but not miss essential information.
Queuing	Delaying response during high load on the assumption that it will be possible to catch up later; (stacking input).	If it is important not to miss any information (this is efficient only for temporary conditions).
Filtering	Neglecting to process certain categories; non-processed information is lost.	If it time/capacity restrictions are really severe and it is sufficient to note only large variations.
Cutting categories	Reduce level of discrimination; use fewer grades or categories to describe input.	
Decentralisation	Distributing processing if possible; calling in assistance.	Availability of additional resources.
Escape	Abandoning the task; giving up completely; leaving the field.	System self-preservation.

In conditions of information input overload, one essential difference between humans and technological artefacts is that humans can degrade gracefully when there is too much information, since they are able to respond in a flexible manner as described by the strategies listed in Table 4.1. Humans will try to do as well as possible, for as long as possible. Artefacts, on the other hand, tend abruptly to botch up. This difference is also found for conditions of information underload, cf. below.

Information Input Underload

If too much information clearly can be a problem, so can too little. In this case, which we may call information input underload – and which Miller (1960) only considered as the extreme case of sensory deprivation – there is less information than needed. This means that the JCS has to work from a limited amount of information both in interpreting the situation and in deciding what to do (Reason, 1988).

The problem of information input underload has received even less attention than information input overload (Table 4.2). One possible reason is that overload conditions are more conspicuous, and that overload also is a more striking condition than underload. From a technical point of view it is a problem only if channel or processing capacity is exceeded (Moray, 1967); if the load is less than the capacity, the situation is by definition considered as normal. That is, however, correct from only a quantitative point of view. From a qualitative point of view, information input underload matters because it negatively affects the operator's ability to be in control. Information input underload may occur either if information is missing (a true underload condition) or if it has been discarded for one reason or another, for instance in response to a preceding overload condition.

Table 4.2: Coping Strategies for Information Input Underload

IIU strategy	Definition
Extrapolation	Existing evidence is 'stretched' to fit a new situation; extrapolation is usually linear, and is often based on fallacious causal reasoning.
Frequency gambling	The frequency of occurrence of past items/events is used as a basis for recognition/selection
Similarity matching	The subjective similarity of past to present items/events is used as a basis for recognition/selection
Trial-and-error (random selection)	Interpretations and/or selections do not follow any systematic principle.
Laissez-faire	An independent strategy is given up in lieu of just doing what others do.

Whereas coping strategies here have been described as responses to input conditions of either overload or underload, it is more in line with CSE to see them as typical ways of responding and not just as determined by the working conditions. The choice of a coping strategy not only represents a short term or temporary adjustment but may equally well indicate a more permanent style. One example of that is the balance between intuition and analysis mentioned above; another is a generic efficiency-thoroughness trade-off (ETTO; cf., Hollnagel, 2004). Indeed, many of the judgment heuristics described by Tversky & Kahneman (1974), such as representativeness, availability, adjustment and anchoring, or even the concept formation strategies described by Bruner, Goodnow & Austin (1956), such as focus gambling, conservative focusing, simultaneous scanning, and successive scanning, represent not only temporary adjustments to save the situation but also long-term strategies that are used for many different situations. Coping with complexity in a long-term perspective may require a JCS to conserve effort and keep spare capacity for an 'emergency', hence to make a trade-off before it is objectively required by the current situation.

DESIGNING FOR SIMPLICITY

The growing system complexity can lead to a mismatch between (task) demand and (controller) capacity. This mismatch can in principle be reduced either by reducing the demands or by increasing the capacity – or by doing both at the same time. As described in Chapter 2, the history of technology can be seen as a history of amplification of human capabilities, directly or indirectly. In our time this has opened up a world of decision support systems, pilot's associates, human-centred automation, and so on, of which more will be said in Chapter 6. This has been a major area of research and development since the early 1980s and remains so despite the often disappointing results.

Another way to extend capacity is to select people who excel in some way or other. Selection can be enhanced by training people to ensure that they have the necessary competence to accomplish the required tasks. An extreme example of that is the training of astronauts. More down to earth is the training of operators for complex processes, such as nuclear power plants or fighter aircraft, where the training may take years. Since training, however, is costly and furthermore must be maintained, considerable efforts are put into making systems easier to operate, for instance by increased automation and/or better design of the work environment, thereby reducing the demands for specialised training.

A reduction of the mismatch can also be achieved by reducing the demand by smart system design, specifically by simplifying the information presentation. This approach has been taken to the extreme in the pursuit of user-friendly products of all kinds. Here the ideal is completely to eliminate the need to cope with complexity by making the use of the artefact 'intuitive'. This means that everyone should be able to use the artefact immediately, or at least after reading the 'quick start' instructions that today accompany most consumer products with some modicum of functionality. This has led to the adage of 'designing for simplicity', which, if taken seriously, means that the complexity becomes hidden behind the exterior, so that the requisite variety is reduced to something that corresponds to the 'innate' abilities of humans. By putting it this way it is easy to see that this is an approach doomed to failure, for the reason that complexity or variety is not actually reduced but rather increased, as the entropy of the modified artefact is increased. Another problem is that there is little or no consensus about how the 'innate' or minimal abilities of humans should be determined.

Simplicity-Complexity Trade-Off

In discussing the trade-off between simplicity and complexity there are a number of inconvenient facts that must be recognised. The first is that both complexity and simplicity are epistemological rather than ontological qualities, as discussed above. This means that the degree of complexity or simplicity of the description of a system is relative to the user, to the point of view taken, and to the context and purpose of the description. This relativity means that any description is vulnerable to the $n+1$ fallacy. The $n+1$ fallacy refers to the fact that while it is possible to describe the system for n different conditions, hence to account for the complexity under these conditions, there will always be a condition that has not been accounted for, which is the $n+1$ condition. This is so regardless of how large n is. The consequence for system design is that it cannot be guaranteed that the system description will be simple also for the $n+1$ situation; the very principle of designing for simplicity therefore has a built-in limitation.

Any local improvement will invariably be offset by an increase in complexity and variety of other situations – essentially those that have not been included in the design base – and will therefore lead to an overall increase in complexity. This may not have any discernible effects for long periods of time, but nevertheless remains a potential risk, similar to the notion of latent conditions in epidemiological accident theories (Reason, 1997). In this case the risk is unfortunately one that has been brought into the system by well-meaning interface designers, rather than one, which has occurred haphazardly or unexpectedly.

Information Structuring

The common basis for interface design has been the 'right-right-right' rule, according to which the solution is to display or present the right information, in the right form, and at the right time. This design principle is simplicity itself and would undoubtedly have a significant effect on practice if it were only possible to realise it. The basic rationale for the design principle is the hindsight bias to which we all succumb from time to time – although some do so more often than others. The essence of that bias is that when we analyse an incident or accident that has happened, such as the Apollo 13 problem (Woods, 1995), we can very easily propose one or several solutions that would have avoided the problem, if only they had been implemented in time. (The hindsight bias is thus equivalent to the fallacy of relying on counterfactual conditionals.) That is, if only information X had been presented in format Y, then the poor operators caught by unfortunate circumstances would have understood the situation correctly, hence not have failed. The shortcomings of reasoning in this manner are, however, easily exposed if we look at each of the elements of the 'right-right-right' rule in turn.

The Right Information

The right information can be determined in two principally different ways. One is by referring to a specific – and preferably 'strong' – theory of human action. In practice, this has often been a theory of human information processing, which – emphatically – is not the same as a theory of human action. The literature is awash with various display design principles, although some are more principled than others. One early example is the work of Goodstein (1981), Rouse (1981) and, of course, Rasmussen & Vicente (1987) (see also Vicente & Rasmussen, 1992). The latter is representative of what we talk about here, because the display design refers to one of the better-known version of a human information-processing model.

Another way of determining what the right information is in advance is to consider specific situations as they are defined by the system design. An almost trivial example of that is a situation described by operating procedures – both for normal operations and emergencies. In these cases it is clearly possible from the procedure to derive what the information demands are – both in content and in form. Assuming that the operators follow the procedures rigidly, and also that the procedures are valid for the situation, the display problem is in principle solvable. Examples of that are found in task-based information displays (O'Hara et al., 2002) and in the various structuring of procedures (event-based, symptom-based, critical function based), e.g., Colquhoun (1984).

A less stringent criterion is to identify the information that is typically required for a range of situations, and which can be derived from, e.g., a control theoretic analysis of the system (Lind & Larsen, 1995). An example that has been widely used by industry is the so-called star display developed in the 1980s. Another example, which also has considerable practical merit, is the concept of critical function monitoring (Corcoran et al., 1981). Similar examples may be found in aviation, where much ingenuity and creativity go into the design of the EDIC precisely because it is a very limited display area hence enforces a keyhole effect (Woods & Watts, 1997).

In the Right Form

Assuming that the problem of determining what the right information is has been solved, the next challenge is to present the information in the right form. For any specific information contents there is a large variety of information forms. The importance of form (or structure) has been studied in problem solving psychology at least since Duncker (1945) and is an essential issue in information presentation and display design (Tufte, 2001).

Most of the proposed design principles do in fact combine content and structure (form), although it is always possible to suggest alternative ways of structuring a given set of data. This, however, is precisely the problem since it is very difficult to determine in advance what the best possible representation format is. A great number of factors must be taken into consideration, such as personal preferences, experience, norms and standards of the work environment, whether work is by a single user or a team, etc. Other factors are the temporal and task demands, e.g. whether there are multiple tasks and severe time constraints or few tasks and a more leisurely pace. Clearly, in a state of high alert and with a need to respond quickly, it may be advantageous either to highlight the important piece of information or to filter out less important information. Both solutions affect the form of the presentation, but neither is easy to do in a consistent manner.

Since it is impossible to reduce the *real complexity*, the alternative solution is to reduce the *perceived complexity* of the system by simplifying the information presentation. The reasoning is that if the system can be made to *look* simpler, then it will also *be* simpler to control. The fundamental problem with this approach is that it shifts the complexity from the exterior to the interior of the system. Designing for simplicity does not actually reduce complexity or eliminate demands but only changes their appearance and focus. That the principle nevertheless has had a considerable degree of success, for instance as ecological interface design (Vicente & Rasmussen, 1992), is due to the simple fact that *the effects of complexity are unevenly distributed over all possible situations*. Good interface or good interaction design may in many cases produce a local reduction in complexity, for instance for commonly occurring situations or conditions, which means that these tasks become easier to accomplish. What should not be forgotten is that this advantage has a price, namely that the very same displays may be less optimal – or downright inconvenient – in other situations. Just as there is no such thing as a universal tool, there is no such thing as a universal display format.

Technically, the problem is often expressed as a question of identifying user needs and preferences, with suggestions that the problem is solved by finding or having the right user model (e.g. Rich, 1983). The concept of a user model is, however, close to being conceptually vacuous, quite apart from the fact that it reiterates the unfortunate separateness between the user on one side and the machine on the other (cf. Chapter 3). User models are not a viable solution as variability within and between situations and users is too large. Focusing on the issue of information contents and information structure also tacitly accepts that the problem is one of transmitting information from the interface to the user, rather than one of ensuring that the JCS can maintain control. Even if information was presented so that it could be understood with little effort and no ambiguity, there would be no guarantee that a person would be able to find the right response and effectuate it in time. The issue of information presentation puts too much weight on the input and too little on the output; i.e., it considers only one side of the coin, so to speak.

At the Right Time

Presenting the information *at the right time* is the final challenge. Fortunately, this problem is a little simpler to solve than the previous ones because timing can be defined in relation to the process state, which in many cases is known or at least identifiable.

The question of timing can be seen as a question of presenting the information neither too early, nor too late. One criterion could be that certain conditions obtain that define the right time. For instance, in a collision detection system for air traffic management, the indication of a possible conflict can be defined relative to a given criterion (the separation distance). For most industrial processes, alarms are shown only when given conditions are present. These conditions, such as a high level in a tank, define when the information is needed, at least from the process' point of view. In practice this may be either before or after it is needed by the operator.

The problem of whether the information is presented too late can be solved in a similar manner, since the cessation of a process state or process condition can be used as an indication. This, again, is a frequently applied principle in presenting alarm information, where many alarms automatically are cancelled by the system when the defining condition disappears.

The problem of the timing of information presentation is more difficult if the criterion is the operator's needs in the situation. In practice it is nearly impossible to determine when an operator needs information of a certain type (and in a certain form), since there is no way of determining the operator's mental state. There have nevertheless been several proposals for adaptive interfaces and adaptive information presentation (e.g., Onken & Feraric, 1997), as well as more specific studies of the triggering conditions for such systems (Vanderhaegen et al., 1994). An alternative to concentrating on the operator is to use a combination of system state and operator state to determine the timing issue, and furthermore to let the determination of the operator's state depend on measurable performance indicators rather than inferred mental states (Alty, Khalil & Vianno, 2001).

How Should the Interaction Be designed?

The discussion above leads to the conclusions that the principle of designing for simplicity is impractical. It has in a number of cases met with some success, but only when it has been possible to map the conditions or scenarios onto a limited number of categories or views. Furthermore, it must be possible to make some kind of transformation or projection from the situations onto the categories or views. In general, the principle of designing for simplicity requires that situations can be decomposed and described structurally, and also that a sufficiently strong principle or theory of how humans make use of that information is available. As we have shown repeatedly in this book, there are reasons to be wary about the decomposition principle, since it provides a conceptually simpler world at the cost of a reduced match to reality.

DESIGNING FOR COMPLEXITY

The logical alternative to designing for simplicity is to design for complexity. This may at first sound counterintuitive, since complexity was the initial problem. Designing for complexity nevertheless makes sense because it recognises that epistemological complexity cannot be reduced to an arbitrary low level, which means that it can never be simplified so much that it can be ignored. In consonance with the principles of CSE, the goal of 'designing for complexity' is therefore to enable the JCS to retain control, rather than to simplify the interaction or the interface.

Designing for complexity is also in good agreement with the Law of Requisite Variety. Since the controller of a system must have at least as much variety as the system to be controlled and since the requisite variety cannot be reduced by interface design alone, the design goal must be to enhance or increase the variety of the controller. This cannot be achieved by hiding complexity, but possibly by making it visible, thereby allowing the JCS to learn and increase its own variability. Designing for complexity reflects a basic principle of CSE that the ability to maintain control depends on the ability to understand what happens and to predict what will happen. Referring to Figure 4.3 above, coping is facilitated if the JCS knows what has happened and what happens, and if it can predict likely developments and thereby anticipate the effect of possible actions (i.e., the classical principles of observability and controllability).

Designing for complexity fully acknowledges that the work environment of the operator is complex but draws the conclusion that this complexity is necessary for effective control. Designing for complexity is not hostile to the use of technology. On the contrary, the foundation for the principle is the recognition that the *joint* cognitive system is a fundamental entity. The difference is that designing for complexity aims to support general functions of coping, rather than specific ways of acting in particular situations, hence chooses generality over specificity. Some of the main principles that can be applied to that are described next.

Support for Coping

One principle is that the design should support the natural human strategies for coping, rather than enforce a particular strategy. A trivial example is to consider the two extreme views of decisions making, rational decision making and naturalistic decision making. Rational decision making is based on a set of strong assumptions of what decision making is and what the human abilities of being a rational decision maker are (Petersen & Beach, 1967). On the basis of that, it is possible to propose and build tools and environments that support decision making – but only as it is described by the model. For naturalistic decision making, and more generally for what are called the ecological or ethnographic approaches to human performance, the method would be to study what people actually do and then consider whether it is possible to support that through design (e.g., Hutchins, 1995). This approach is not atheoretical, but the theories are about what people do rather than about the hypothetical 'mechanisms' behind. (An excellent example of this approach is provided by Neisser (1982) – a decade before 'cognition in the wild' became a catchphrase.)

The starting point, in other words, should be in an understanding of the representative strategies for coping, with no initial assumptions about what goes on in the operator's mind. That understanding must obviously be established separately for each domain and field of practice. It is unreasonable to assume that there are strong domain-independent practices, and that design can be based exclusively on those. Having said that, it is to be expected that significant common features exist among domains, although these may emanate from the work demands as likely as from inherent psychological characteristics. In many cases there will be a need to control or focus attention on important facets, for instance by applying a good representation, cf. Gibson's (1979) notion of perceived affordance. As discussed above it is nevertheless far from easy to determine *a priori* what will be significant and what will not. In general, since time is limited – and time is perhaps the only true common feature across all situations – there will be an advantage in reducing, filtering, and transforming information to avoid obvious performance bottlenecks.

As a concrete illustration, at least three of the common strategies of coping with input information overload – queuing, filtering, and cutting categories – can be supported by interface design, and, in fact, often are, although probably by coincidence rather than by design. Queuing is a feature of VDU-based alarm systems, but may be used more systematically (Niwa & Hollnagel, 2001). Filtering can be supported by categorising plant data and measurements, for instance, according to urgency. And cutting categories can be done by algorithmically mapping complex measurements onto a limited set of more abstract functions (Corcoran et al., 1981).

While designing for complexity neither can nor should advocate a single principle or paradigm, it does provide a number of general principles, design rules or heuristics that should be kept in mind. If these are followed, the result should be a JCS that is better able to handle a wide range of situations. The disadvantage is that the principles are high-level guidelines rather than low-level design rules. They thus require a certain amount of interpretation as well as some experience in system design. But this is only to be expected since one cannot design a complex system or a complex interface without knowing something about what lies behind. Notwithstanding many promises, there are no simple step-by-step rules (like paint-by-numbers) that will allow a person without sufficient experience to design a good system. There are many sets of rules that can be used to check the outcome of a design and to evaluate a system, but there are none rules that will produce it in the fist place.

Time

An important principle in designing for complexity is to provide sufficient time for the JCS to do its work. This is in most cases easier said than done, since it is usually very difficult – if not impossible – to reduce the speed of a process sufficiently. Physical processes and chemical reactions, for instance, have their natural rate of development according to the laws of nature, and an airplane must fly fast enough to create lift. Delaying or slowing down a process is possible only in rare cases, such as stock market trading where there may be a built-in 'freeze' if the volume of trading passes a given limit. In cases where the speed of the process cannot itself be reduced, it may sometimes be possible to buy some time by providing additional resources for the control functions, i.e., increase the speed of the controller. In terms of the cyclical model that means reducing the time needed for evaluation and selection by introducing parallel processes, for instance by calling in additional staff or in other ways lightening the task load.

Slowing down is in practice feasible only for processes where the user is in direct control of the speed, such as driving a car. In this case there are no physical limitations on the lowest possible speed, unlike, e.g., flying, although the surrounding traffic may provide an obstacle. It is, indeed, a common approach to pull up by the curb when orientation has been lost, while driving in an unfamiliar environment – whether in a city or in the countryside. Typically (in a city) the first reaction to losing orientation is to slow down, to enable street names to be read and landmarks to be recognised (depending on whether one is alone in the car or has a map-reading passenger to assist). If this fails, the ultimate option is to stop completely and spend enough time to re-establish orientation. Note that driving in an unfamiliar environment also is a good example of how planning is interwoven with actions, since the planning serves to identify the specific marks or waypoints that guide the actual driving.

An interesting example of providing more time is the so-called 30-minute rule that exists for nuclear power plants. According to this rule, the automatic safety systems of the plant must be able to keep the reactor under control for a period of 30 minutes, thereby giving the operators time to think. In the strict interpretation of this rule, operators are not required – or even allowed – to respond for the first 30 minutes. This rule does not slow down the process as such, as the nuclear reaction continues at its own pace, but it does provide the time needed to assess the situation and decide on a response. (There is apparently no solid scientific or empirical reason for setting the limit to 30 minutes. In Japan, for instance, the corresponding rule calls for a 10-minute respite.)

Another, but more indirect, way to provide sufficient time is to ensure that the information presentation and the interface are as easy to use as possible. An effective structuring of the information presentation will reduce the demands to work. Conversely, a poor structuring of the information and an inconsistent design of how the interface is controlled may lead to unnecessary secondary tasks.

Predictability

Another way in which coping can be enhanced is by providing good predictions or by supporting anticipation, either explicitly or implicitly. Explicit support means that actual predictions are provided by some kind of technology, raging from simple graphical extrapolations of trends (so-called Janus displays, named after the Roman god of gates and doorways who had two faces looking in opposite directions), over calculations of projected developments such as in aviation and sailing (the position five minutes hence), to faster than real-time simulations of not only the system itself but also the environment. Well-known examples are the Traffic Alert & Collision Avoidance System (TCAS) used in aviation, weather forecasts, warnings of hurricanes and tsunamis, earthquake predictions (which usually have limited success), market forecasts (also with limited success), predictions of greenhouse effects, etc. On the fringe, the use of horoscopes and psychic readings also illustrates the inexorable need for predictions, although the accuracy may leave something to be desired.

Predictions are often partially supported by the way in which information is presented. A simple example is when the level in a tank (or the value of a stock or a currency) is shown graphically instead of digitally, i.e., by showing present and past values rather than just the present value. Although display design generally has been used to support designing for simplicity, many of the established techniques can equally well be used to support designing for complexity. It is generally beneficial if it becomes easier for the user to anticipate what may happen, in the sense of being able to see what the outcome of specific actions will be. Predictions can thus be aimed at the future states of the process as it develops according to its own principles or as a result of specific control actions.

Coping can also be enhanced by providing clear alternatives, for evaluation but in particular for planning and the selection of the next action. This is actually the basic reason for introducing decision support and procedures. Decision support provides those alternatives for action that are appropriate in the given situation, and may be combined with predictive support to provide people with realistic expectations of what may happen.

Procedures are interesting since they both reduce task demands by alleviating requirements to memory and help decision making by providing a prepared description of what the alternative possibilities for actions are. (This goes for conventional as well as for computerised procedures.) The difference is between finding out oneself what shall be done and perhaps even plan and schedule it in detail on the one hand, and following the instructions or procedures on the other. Procedures are a really important way of coping with complexity, since the complexity in a sense becomes encapsulated in the procedure. The downside is, of course, that the procedure and the conditions must match to a very high degree, i.e., that the procedure must be appropriate for the conditions. This is often a serious stumbling block; if there is any significant discrepancy the procedure does not support coping but may on the contrary lead to a loss of control by recommending or enforcing inappropriate actions by an unsuspecting operator.

Summary

Both designing for simplicity and designing for complexity aim to make it easier for an operator to control a process. The two approaches differ in how they go about doing that, because of the basic difference between how they view human-machine interaction and work.

Designing for simplicity is possible only if the complexity of the real world can be transformed according to a set of well defined, and valid, principles. Experience has demonstrated in a quite unequivocal fashion that this is rarely the case. Therefore, rather than designing for a simple world that does not exist, the goal should be to design for the complex world that does. This may require a serious rethinking of current approaches, and probably the discarding of a few sacred beliefs. It is, however, becoming clear that such a change is necessary to break the current impasse.

Chapter 5

Use of Artefacts

Humans have used artefacts since the invention of cuneiform writing, but the complexity of present-day computerised artefacts often create more problems than they solve. Artefacts can be either tools or prostheses, and the chapter develops these distinctions. While artefacts (technical or social) ostensibly are introduced to make life easier for people, the effect sometimes turn out to be the exact opposite.

INTRODUCTION

Much has been written about the use of tools, in particular the type of tools that include computers and information technology. Indeed, in discussions in the human factors/human-machine interaction communities since the early 1980s, the notion of a tool has practically become synonymous with a computer or something that includes a computer. An illustration of this is Bødker's (1996) proposal for a distinction between tool, medium, and system. According to this, a tool emphasises human involvement with materials through a computer application, a medium emphasises human involvement with other human beings through the computer, and a system emphasises the perspectives of the human user and data exchange with the compute component. Although the importance of computers for present day society is hard to underestimate there still are many artefacts that are not computers or that do not involve computers, but which nevertheless are worthy of consideration.

Since CSE uses the concept of a tool in a more restrictive sense than above, the preferred term is an *artefact*, defined as something made for a specific purpose. Humans sometimes use natural objects to achieve a goal, such as when a stone serves as a hammer, and sometimes use artefacts for purposes they were not made for, e.g., using a fork as a lever to open a bottle of beer. Depending on how an artefact is used, it may be considered either a tool or a prosthesis. Computers, and systems including computers, are certainly artefacts in this sense, but so are bicycles, sewing machines, keys, aeroplanes, the Internet, personal digital assistants (PDA), dishwashers, cars, photocopiers – the list is endless. Although CSE does not aim to exclude any artefact, the emphasis is nevertheless on those that have a certain level of complexity and functionality and which therefore typically – but not necessarily – comprise some kind of information technology. The interest is, however, not on the information technology as such but on the function and use of the artefact.

Phenomenology of Coagency

The extent to which people in the industrialised world have become dependent on artefacts is easily demonstrated by considering an ordinary day of work. We wake up in the morning by the sound from an alarm clock. We go to the bathroom to wash, brush our teeth, and perhaps shave. We put on clothes and go to the kitchen where breakfast is prepared using microwaves, stoves, electric kettles, toasters, and the like. We may listen to the radio, watch the news, or read the paper. We travel to work equipped with various technological artefacts such as watches, mobile phones, computers, music players or radios, and journey by means of bicycles, cars, trains, boats, and buses – except for the fortunate few who live within walking distance. Even so, no one walks barefoot and naked to work. At work we use a variety of artefacts and machines and are furthermore surrounded by artefacts in the form of furniture, lighting, heating, ventilation, and perhaps even monitored by cameras, movement detectors, and the like – there is hardly any need to continue. Indeed, it is difficult to think of a single thing that can be done without the use of some kind of artefact, with the possible exception of a vegan nudist living in the woods.

Philosophers, such as Edmund Husserl and Martin Heidegger, have provided the basis for a phenomenology of human-machine relations – of coagency (Ihde, 1979, p. 3). The basis is a distinction between the embodiment and hermeneutic relations, described in Chapter 2. In the embodiment relation, the artefact or machine becomes transparent to the user so that it is no longer seen or experienced as an object but instead becomes part of how the world is experienced. A simple example is to write with a pencil on a piece of paper. In the act of writing the pencil stops being an object and becomes, for the moment at least, at one with or an embodiment of the writer. This can also be expressed by saying that there is a transparency relation between the artefact and the user. The better the artefact is suited for its purpose, the higher is the transparency. The embodiment relation can be illustrated as in Figure 5.1. The shaded area indicates that the artefact or machine is transparent to the person, and that the interface is between the artefact and the world, rather than between the user and the artefact.

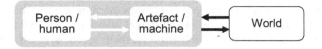

Figure 5.1: The embodiment relation.

In the embodiment relation the artefact often serves as an amplifier, i.e., to strengthen some human capability. The amplification highlights those aspects of the experience that are germane to the task while simultaneously reducing or excluding others, all – ideally – controlled by the user. In the hermeneutic relation, the artefact stands between the person and the world. Instead of experiencing the world through the artefact, the user experiences the artefact, which thereby interprets the world for the person.

The hermeneutic relation can be illustrated as in Figure 5.2. The shaded area indicates that the world is experienced only as the artefact represents it. The interface is now between the person and the artefact, rather than between the artefact and the world. (The similarity to the traditional HCI paradigm described in Chapter 1 is striking.)

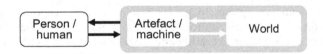

Figure 5.2: The hermeneutic relation.

In the hermeneutic relation, the artefact serves as an interpreter for the user and effectively takes care of all communication between the operator and the application. By virtue of that, the artefact loses its transparency and becomes something that the user must go through to get to the world. The interaction in many cases therefore becomes with the artefact rather than with the application. Put differently, the user has moved from an experience *through* the artefact to an experience *of* the artefact. In the extreme case there actually is no experience of the process except as provided by the artefact, which therefore serves as a mediator without the user's control.

Tools and Prostheses

The embodiment and hermeneutic relations are, of course, not mutually exclusive but rather represent two different ways of viewing the human-machine ensemble. The distinction between them serves to clarify the nature of JCSs. It also closely resembles the distinction between using technology as a tool or as prosthesis. (Tool is a word found in Old English meaning instrument or implement. Prosthesis comes from Greek and originally means the addition of a letter or a syllable to a word. It was later defined as an artificial substitute for a missing part of the body, and in the present context it means an artefact that takes over an existing function.)

In an insightful analysis of the problems that seemed to be endemic to work with complex processes, Reason (1988) presented what he termed an optimistic and a pessimistic view. According to the optimistic view, technology would bring the solution to its own problems by providing cognitive tools – described as "felicitous extensions of normal brain-power" (Reason, 1988, p. 7) – that would enable people to cope with the complexity and opacity of industrial systems. According to the pessimistic view, the more likely remedy would be to provide people with cognitive prostheses (or mental 'crutches') that would help to compensate for some of the 'error' tendencies that have their root in a mismatch between the properties of the system as a whole and the characteristics of human information processing. If this distinction is extended from cognitive or computer-based artefacts to all types of artefacts, a tool is an artefact (in the widest possible sense) that enables a cognitive system to go beyond its unaided capabilities without losing control. Conversely, a prosthesis is an artefact that enables a cognitive system to do things that were hitherto impossible, but which takes control away – in whole or in part.

The tool-prosthesis distinction is not absolute and in practice the two terms are not mutually exclusive. Indeed, any given artefact may have both tool-like and prosthesis-like qualities, meaning that it sometimes functions as a tool and sometimes as a prosthesis. Furthermore, the function as either a tool or a prosthesis may change as the user becomes more experienced. For the skilled user an artefact may be a tool that can be used in a flexible way to achieve new goals. For the unskilled or unpractised user, the very same artefact may be as a prosthesis – something that is necessary to achieve a goal, but which also makes control more difficult. The artefact works, but the user does not really understand why and has few possibilities of influencing it.

Artificial intelligence (AI), as an example of a specialised technological development, clearly provides the possibility of amplifying the user's capabilities, hence to serve as a tool. Yet AI has often been used to enhance the computer as an interpreter, thereby strengthening its role as a prosthesis.

This may happen as a simple consequence of the increasing complexity of the applications and of the perceived need to provide access to compiled expertise. The very development of expert systems may thus unknowingly favour the dominance of the role of the computer as an interpreter (Weir & Alty, 1989). The development of user support systems and advanced human-machine system functions should, however, rather aim at amplifying the capacities of the user.

The tool relation of human-machine interaction means that the computer is used to strengthen coagency rather than to reduce the environment until it fits the inherent capacity of putative human information processing mechanisms. An example would be the use of computers in a process control environment, such as a chemical plant or a hospital. Here the computer may assist the operator in monitoring, detection, sequencing, compression, planning/scheduling, diagnosing, etc. Computers and automation should only take over these functions to the extent that this can be proven to further the overall task.

One solution would be to introduce support that is goal directed, i.e., where the system is given a goal and then itself finds a way of achieving it. This will also ensure that the operator remains in control and would differ radically from designing systems that contain the (top level) goals themselves, and enforce them on the operator. A system that works by being given goals corresponds to the delegation of functions that can be observed in good organisations and in good human cooperation. People are on the whole better at defining goals than at specifying procedures and better at achieving goals than at following instructions to the letter. Goal definition requires creativity and comprehension, the ability to see patterns, etc., all of which computers are bad at. Generating detailed procedures (prescribing the solutions to the goals) require massive computations and a meticulous following of rules, which people notably are bad at.

ARTEFACTS IN CSE

The distinctions between an embodiment and a hermeneutic relation on the one hand, and a tool and prosthesis perspective on the other, are not quite identical because they depend on each other. One way of describing the relationship is as shown in Figure 5.3, which uses the two dimensions of transparency and exchangeability. A high degree of transparency corresponds to an embodiment relation and a low to a hermeneutic relation. Similarly, a high degree of exchangeability corresponds to a prosthetic relation (the artefacts functions as a prosthesis that can easily be replaced), while a low corresponds to a tool relation (i.e., the artefact functions as a tool that amplifies the user's abilities, thereby becoming an integral part of work).

In Figure 5.3, four characteristic artefacts are used to illustrate the meaning of the dimensions. A decision support system, and in general any system that provides a high level of complex functionality in an automated form, illustrates a hermeneutic relation and a high degree of exchangeability – i.e., a prosthetic relation. While decision support systems (e.g., Hollnagel, Mancini & Woods, 1986) are supposed to help users with making decisions, they often carry out so much processing of the incoming data that the user has little way of knowing what is going on. In practice, the user's decision making is exchanged or substituted by that of the DSS, although this may not have been the intention from the start.

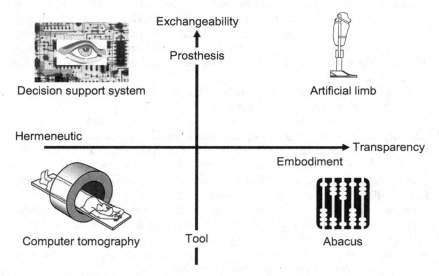

Figure 5.3: The transparency-exchangeability relation.

An artificial limb, such as a prosthetic arm or leg, represents an artefact, which on the one hand clearly is a prosthesis (it takes over a function from the human), but at the same time also is transparent. The user is in control of what the prosthesis does, even though the functionality need not be simple. (A counterexample is, of course, the ill-fated Dr. Strangelove.) The issue becomes critical when we consider a visual prosthesis that allows a blind person to see, at least in a rudimentary fashion. Such a device certainly produces a lot of interpretation, yet it does not – or is not supposed to – provide any distortion. Indeed, it is designed to be transparent to the user, who should 'see' the world and not the artefact.

Computerized tomography (CT), or computerized axial tomography (CAT), as well as other forms of computer-enhanced sensing, illustrate artefacts with a hermeneutical relation, yet as a tool, i.e., with low exchangeability. Computer tomography is a good example of extending the capability of humans, to do things that were hitherto impossible. CAT, as many other techniques of graphical visualisation – including virtual reality – are seductive because they provide a seemingly highly realistic or real picture of something. In the case of CAT there is the comfort that we know from studies of anatomy that the underlying structure actually looks as seen from the picture. In the case of research on subatomic particles – such as the hunt for the Higgs' boson – or in advanced visualisation of, e.g., simulation results and finite element analysis, the danger is considerably larger, hence the hermeneutic relation stronger, because there is no reality that can be inspected independently of the rendering by the artefact. Indeed, one could argue that the perceived reality is an artefact of the theories and methods.

The final example is an abacus. The abacus is highly transparent and for the skilled user enters into an embodiment relation. The human calculator does not manipulate the abacus, but calculates with it almost as a sophisticated extension of hand and brain. Similarly, the abacus is clearly a tool, because it amplifies rather than substitutes what the user can do. The human calculator is never out of control, and the abacus – beings so simple – never does anything that the user does not.

Range of Artefacts

As the example of getting up in the morning showed we are immersed in a world of artefacts, some of which are our own choosing while others are not. For CSE it is important to understand how artefacts are used and how JCSs emerge. For this purpose it is useful to consider several ranges of artefacts: simple, medium, and complex.

The simple artefacts are things such as overhead projectors, household machines, doors, elevators, watches, telephones, calculators, and games. They are epistemologically simple in the sense that they usually have only few parts (considering integrated circuits as single parts in themselves), and that their functions are quite straightforward to understand, hence to operate. (Newer types of mobile phones and more advanced household machines may arguably not belong to this category, but rather to the next.) This means that they are tools rather than prostheses in the sense that people can use them without difficulty and without feeling lost or uncertain about what is going to happen. Indeed, we can use this as a criterion for what should be called a simple artefact, meaning that it is simple in the way it is used, typically having only a single mode of operation, rather than simple in its construction and functionality. (Few of the above mentioned artefacts are simple in the latter sense.) Indeed, much of the effort in the design of artefacts is aimed at making them simple to use, as we shall discuss below.

Medium level artefacts can still be used by a single person, i.e., they do not require a co-ordinated effort of several people, but may require several steps or actions to be operated. Examples are sports bikes (mountain bikes in contrast to ordinary city bikes), radios, home tools, ATMs – and with them many machines in the public domain, VCRs, personal computers regardless of size, high-end household machines, etc. These artefacts are characterised by having multiple modes of operation rather than a single mode and in some cases by comprising or harnessing a process or providing access to a process. They may also comprise simple types of automation in subsystems or the ability to execute stored programs. In most cases these artefacts are still tools, but there are cases where they become prostheses as their degree of exchangeability increases.

Finally, complex artefacts usually harbour one or more processes, considerable amounts of automation, and quite complex functionality with substantial demands to control. Examples are cars, industrial tools and numerically controlled machines, scientific instruments, aeroplanes, industrial processes, computer support systems, simulations, etc. These artefacts are complex and more often prostheses than tools, in the sense that the transparency certainly is low. Most of them are associated with work rather than personal use, and for that reason simplicity of functioning is less important than efficacy. Usually one can count on the operator having some kind of specific training or education to use the artefact, although it may become a prosthesis nevertheless.

Cognitive Artefacts

It has become common in recent years to refer to what is called cognitive artefacts. According to Norman (1993), cognitive artefacts are physical objects made by humans for the purpose of aiding, enhancing, or improving cognition. Examples are sticker-notes or a string around the finger to help us remember. More elaborate definitions point to artefacts that directly help cognition by supporting it or substituting for it.

Strictly speaking, the artefacts are usually not cognitive – or cognitive systems – as such, although that is by no means precluded. (The reason for the imprecise use of the term is probably due to the widespread habit of talking about 'cognitive models' when what really is meant is 'models of cognition' – where such models can be expressed in a number of different ways.) There is indeed nothing about an artefact that is unique for its use to support or amplify cognition. A 5-7000 year old clay tablet with cuneiform writing is as much an artefact to support cognition as a present day expert system. Taking up the example from Chapter 3, a pair of scissors may be used as a reminder (e.g., to buy fabric) if they are placed on the table in the hall or as a decision aid if they are spun around.

Consider, for instance, the string around the finger. Taken by itself the string is an artefact, which has a number of primary purposes such as to hold things together, to support objects hanging from a point of suspension, to measure distances, etc. When a piece of string is tied around the finger it is not used for any of its primary purposes but for something else, namely a reminder to remember something. Having a string tied around the finger is for most people an unusual condition and furthermore not one that can happen haphazardly. Finding the string around the finger leads to the question of why it is there, which in turn (hopefully) reminds the person that there is something to be remembered. In other words, it is a way to extend control by providing an external cue.

The use of the string around the finger is an example of what we may call *exocognition* or the externalisation of cognition. Rather than trying to remember something directly – by *endocognition* or good old-fashioned 'cognition in the mind' – the process of remembering is externalised. The string in itself does not remember anything at all, but the fact that we use the string in a specific, but atypical way – which furthermore only is effective because it relies on a culturally established practice – makes it possible to see it as a memory token although not as a type of memory itself.

We may try to achieve the same effect by entering an item, such as an appointment, into the to-do list of a personal digital assistant (PDA) or into the agenda. In this case the PDA can generate an alert signal at a predefined time and even provide written reminder of the appointment. Here the PDA is more active than the piece of string, because it actually does the remembering; i.e., it is not just a cue but the function in itself. The PDA is an amplifier of memory, but it only works if there is someone around when the alert is generated. Relative to the PDA, the piece of string around the finger has the advantage of being there under all conditions (until it is removed, that is). Neither the string nor the PDA is, however, cognitive in itself – regardless of whether we use the traditional definition or the CSE interpretation. Speaking more precisely, they are artefacts that amplify the ability to control or to achieve something or which amplifies the user's cognition, rather than cognitive artefacts *per se*. As such they differ in an important manner from artefacts that amplify the user's ability to do physical work, to move, to reach, etc., in that they only indirectly strengthen the function in question.

The Substitution Myth

It is a common myth that artefacts can be value neutral in the sense that the introduction of an artefact into a system only has the intended and no unintended effects. The basis for this myth is the concept of interchangeability as used in the production industry, and as it was the basis for mass production – even before Henry Ford. Thus if we have a number of identical parts, we can replace one part by another without any adverse effects, i.e., without any side-effects.

While this in practice holds for simple artefacts, on the level of nuts and bolts, it does not hold for complex artefacts. (Indeed, it may not even hold for simple artefacts. To replace a worn out part with a new means that the new part must function in a system that itself is worn out. A new part in an aged system may induce strains that the aged system can no longer tolerate.) A complex artefact, which is active rather than passive, i.e., one that requires some kind of interaction either with other artefacts or subsystems, or with users, is never value neutral. In other words, introducing such an artefact in a system will cause changes that may go beyond what was intended and be unwanted (Hollnagel, 2003).

Consider the following simple example: a new photocopier is introduced with the purposes of increasing throughput and reducing costs. While some instruction may be given on how to use it, this is normally something that is left for the users to take care of themselves. (It may conveniently be assumed that the photocopier is well-designed from an ergonomic point of view, hence that it can be used by the general public without specialised instructions. This assumption is unfortunately not always fulfilled.) The first unforeseen – but not unforeseeable – effect is that people need to learn how to use the new machine, both the straightforward functions and the more complicated ones that are less easy to master (such as removing stuck paper). Another and more enduring effect is changes to how the photocopier is used in daily work. For instance, if copying becomes significantly faster and cheaper, people will copy more. This means that they change their daily routines, perhaps that more people use the copier, thereby paradoxically increasing delays and waiting times, etc. Such effects on the organisation of work for individuals, as well as on the distribution of work among people, are widespread, as the following quotation shows:

> New tools alter the tasks for which they were designed, indeed alter the situations in which the tasks occur and even the conditions that cause people to want to engage in the tasks. (Carroll & Campbell, 1988, p. 4)

The substitution myth is valid only in cases where one system element is replaced by another with identical structure and function. Yet even in such cases it is necessary that the rest of the system has not changed over time, hence that the conditions for the functioning of the replaced element are the original ones. This is probably never the case, except for software, but since there would be no reason to replace a failed software module with an identical one, the conclusion remains that the substitution myth is invalid.

CONSEQUENCES OF TECHNOLOGY CHANGE

Since technology is not value neutral, it is necessary to consider in further detail what the consequences of changes may be and how they can be anticipated and taken into account. The introduction of new or improved artefacts means that new tasks will appear and that old tasks will change or even disappear. Indeed, the purpose of design is partly to accommodate such changes either by shaping the interface, by instructions and training, etc. In some cases the changes are palpable and immediate. In many other cases the changes are less obvious and only occur after a while. This is because it takes time for people to adapt to the peculiarities of a new artefact and a changed system, such as in task tailoring. From a cybernetic perspective, humans and machines gradually develop an equilibrium where demands and resources match reasonably well. This equilibrium is partly the result of system tailoring, partly of task tailoring (Cook & Woods, 1996). Technically speaking an equilibrium is said to exist when a state and a transformation are so related that the transformation does not cause the state to change (Ashby, 1956, p. 73-74). From this point of view technology is value neutral only if the new artefact does not disturb the established equilibrium.

One effect of the adjustment and tailoring that makes a system work is that some aspects come to be seen as canonical and while others are seen as exceptional. This means that users come to expect that the system behaves in a specific manner and works in a specific way, i.e., that a range of behaviours are normal. Similarly, there will be a range of infrequent (but actual) behaviours that are seen as abnormal or exceptional, but which people nevertheless come to recognise, and which they therefore are able to handle in some way.

When a system is changed, the canonical and exceptional performances also change. The canonical performance comprises the intended functions and uses, which by and large can be anticipated – indeed that is what design is all about. But the exceptional performance is, by definition, not anticipated, hence will be new to the user. Representative samples of exceptional performance may furthermore take a long time appear, since by definition they are infrequent and unusual. After a given time interval – say six months – people will have developed a new understanding of what is canonical and what is exceptional. The understanding of canonical performance will by and large be correct because this has occurred frequently enough. There is, however, no guarantee that the understanding of exceptional performance will also be correct.

The introduction of a new artefact can be seen as a disruption of an equilibrium that has developed over a period of time, provided that the system has been left to itself, i.e., that no changes have been made. How long this time is depends on a number of things such as the complexity of the system, the frequency of use, the ease of use, etc. After the disruption from a change, the system will eventually reach a new equilibrium, which however may differ somewhat from the former. It is important to take this transition into account, and to be able to anticipate it. Specifically, if only one part of the system is changed (e.g., the direct functions associated to the artefact) but not others, such as the procedures, rules or the conditions for exceptions, then the transition will create problems since the system in a sense is not ready for the changed conditions.

Specifically, the manifestations of failures – or failure modes – will change. The failure modes are the systematic ways in which incorrect functions or events manifest themselves, such as the time delays in a response. New failure modes may also occur, although new failure types are rare. An example is when air traffic management changes from the use of paper flight strips to electronic flight strips. With paper flight strips one strip may get lost, or the order (sequence) of several may inadvertently be changed. With electronic strips, i.e., the flight strips represented on a computer screen rather than on paper, either failure mode becomes impossible. On the other hand, all the strips may become lost at the same time due to a software glitch, or it may be more difficult to keep track of them.

Finally, since the structure of the system changes, so will the failure pathways, i.e., the ways in which unwanted combinations of influences can occur. Failures are, according to the contemporary view, not the direct consequence of causes as assumed by the sequential accident model, but rather the outcome of coincidences that result from the natural variability of human and system performance (Hollnagel, 2004). If failures are seen as the consequence of haphazard combinations of conditions and events, it follows that any change to a system will lead to possible new coincidences.

Traffic Safety

An excellent example of how the substitution principle fails can be found in the domain of traffic safety. Simply put, the assumption is that traffic safety can be improved by providing cars with better brakes. The argument is probably that if the braking capability of a vehicle is improved, and if the drivers continue to drive as before, then there will be fewer collisions, hence increased safety. The false assumption is, of course, that the drivers will continue as before. The fact of the matter is that drivers will notice the change and that this will affect their way of driving, specifically that they will feel safer and therefore driver faster or with less separation to the car in front.

This issue has been debated in relation to the introduction of ABS systems (anti-blocking brakes), and has been explained or expounded in terms of risk homeostasis (e.g., Wilde, 1982). It is, however, interesting to note that a similar discussion took place already in 1938, long before notions of homeostasis had become common. At that time the discussion was couched in terms of field theory (Lewin, 1936; 1951) and was presented in a seminal paper by Gibson & Crook (1938).

Gibson & Crook proposed that driving could be characterised by means of two different fields. The first, called the field of safe travel, consisted at any given moment of "the field of possible paths which the car may take unimpeded" (p. 454). In the language of Kurt Lewin's Field Theory, the field of safe travel had a positive valence, while objects or features with a negative valence determined the boundaries. The field of safe travel was spatial and ever changing relative to the movements of the car, rather than fixed in space. Steering was accordingly defined as "a perceptually governed series of reactions by the driver of such a sort as to keep the car headed into the middle of the field of safe travel" (p. 456).

Gibson & Crook further assumed that within the field of safe travel there was another field, called the minimum stopping zone, defined by the minimum braking distance needed to bring the car to a halt. The minimum stopping zone would therefore be determined by the speed of the car, as well as other conditions such as road surface, weather, etc.

It was further assumed that drivers habitually would tend to maintain a fixed relation between the field of safe travel and the minimum stopping zone, referred to as the field-zone ratio. If the field of safe travel would be reduced in some way, for instance by increasing traffic density, then the minimum stopping zone would also be reduced be a deceleration of the car. The opposite would happen if the field of safe travel was enlarged, for instance if the car came onto an empty highway. The field-zone ratio was also assumed to be affected by the driver's state or priorities, e.g., it would decrease when the driver was in a hurry.

Returning to the above example of improving the brakes of the car, Gibson & Crook made the following cogent observation:

> Except for emergencies, more efficient brakes on an automobile will not in themselves make driving the automobile any safer (sic!). Better brakes will reduce the absolute size of the minimum stopping zone, it is true, but the driver soon learns this new zone and, since it is his field-zone ratio which remains constant, he allows only the same relative margin between field and zone as before. (Gibson & Crook, 1938, p. 458)

The net effect of better brakes is therefore that the driver increases the speed of the car until the familiar field-zone ratio is re-established. In other words, the effect of making a technological change is that the driver's behaviour changes accordingly. In modern language, the field-zone ratio corresponds to the notion of risk homeostasis, since it expresses a constant risk. This is thus a nice illustration of the general consequences of technological change, although a very early one.

Typical User Responses to New Artefacts

As part of designing of an artefact it is important to anticipate how people will respond to it. By this we do not mean how they will try to use it, in accordance with the interface design or the instructions, i.e., the explicit and intentional relations, but rather the implicit relations, those that are generic for most kinds of artefacts rather than specific to individual artefacts. The generic reactions are important because they often remain hidden or obscure. These reactions are rarely considered in the design of the system, although they ought to be.

Prime among the generic reactions is the tailoring that takes place (Cook & Woods, 1996). One type is the *system tailoring*, by means of which users accommodate new functions and structure interfaces. This can also be seen as a reconfiguration or clarification of the interface in order to make it obvious what should be done, to clearly identify measures and movements, and to put controls in the right place. In many sophisticated products the possibility of system tailoring is anticipated and allowance is made for considerable changes. One trivial example is commercial office software, which allows changes to menus and icons, and a general reconfiguration or restructuring of the interface (using the desk top metaphor). Another example is the modern car, which allows for minor adjustments of seats, mirrors, steering wheel, and in some cases even pedals. In more advanced versions this adjustment is done automatically, i.e., the system carries out the tailoring or rather adapts to the user. This can be done in the case of cars because the frame of reference is the anthropometrical characteristics of the driver. In the case of software it would be more difficult to do since the frame of reference is the subjective preferences or cognitive characteristics, something that is far harder to measure.

In many other cases the system tailoring is unplanned, and therefore not done without effort. (Note that even in the case of planned tailoring, it is only possible to do that within the limits of the built-in flexibility of the system. This a driver cannot change from driving in the right side to driving in the left side of the car – although Buckminster Fuller's *Dymaxion* car from 1933 allowed for that.)

The reason for unplanned tailoring is often that the artefact is incomplete or inadequate in some ways; i.e., that it has not been designed as well as it should have. System tailoring can compensate for minor malfunctions, bugs, and oversights. A good example is how people label – or re-label – things. A concrete example is the lifts found in a university building in Manchester, UK (Figure 5.4). It is common to provide a call button by each lift, but for some reason that was not the case here. Consequently, a sign at the right-hand lift instructed people that the call-button was by the left-hand lift, while a sign at the left-hand lift instructed people that a bell-sound indicated that the right-hand lift had arrived.

Another type of tailoring is *task tailoring*. This happens in the case where it is not sufficient to tailor the system, or where system tailoring is impossible. The alternative is then to tailor the tasks, i.e., to change the way in which the system is used so that it becomes usable despite quirks of the interface and functionality. Task tailoring amounts to adjusting the ways things are done to make them feasible, even if it requires going beyond the given instructions. Task tailoring does not go as far as violations, which are not mandated by an inadequate interface but by other things, such as overly complex rules or (from the operator's point of view) unnecessary demands to, e.g., safety. Task tailoring means that people change the way they go about things so that they can achieve their goals. Task tailoring may, for instance, involve neglecting certain error messages, which always appear and seem to have no meaning but which cannot easily be got rid. The solution is to invoke a common heuristic or an efficiency-thoroughness trade-off (ETTO) rule, such as "there is no need to pay attention to that, it does not mean anything" (Hollnagel, 2004). The risks in doing this should be obvious. Task tailoring may also involve the development of new 'error sensitive' strategies, for instance being more cautious when the system is likely to malfunction or break down.

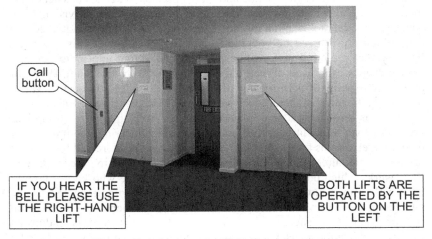

Figure 5.4: Unusual call-button placement.

Failure Modes of an Artefact

In trying to anticipate how something can go wrong, it is useful to consider the failure modes of an artefact. This is a highly developed part of the engineering disciplines that deal with risk and reliability but has for unknown reasons largely been neglected by disciplines such as human factors, human-computer interaction, ergonomics, CSCW, etc. Some attempts of considering failures have been made, notably by Leveson's (1995) work on safeware, but on the whole this practice is rare outside the engineering community.

It is important to understand the failure modes because they may influence whether the artefact is used as a tool or a prosthesis. If something is likely to fail it cannot easily be trusted (Moray et al., 1995). Furthermore, if something can fail it also means that it is not well understood either by the designer or by the user. The fact that failures occur therefore means that we are more likely to objectify the artefact and the use of it, hence distance ourselves from it. The relation will be hermeneutical rather than one of embodiment and the artefact will not be used as smoothly and efficiently as otherwise. It is therefore important not just to design artefacts to do what they are intended to do – and to do that in an easily understandable fashion – but also to ensure either that they do not fail, or that at least the failures are understandable.

It is a fundamental premise of classical ergonomics that interfaces and functionality must be designed specifically to optimise performance in a given task. The ability to describe and model predictable erroneous actions is therefore crucial for good system design. Unfortunately, the provision of an adequate model to explain these cases has proved difficult because psychological processes are both hidden and highly adaptive. This is seen by the way that preferred styles of responding and acting appear to undergo radical change as a result of learning (Hoc, Amalberti & Boreham, 1995), or by how the quality of performance of particular individuals frequently varies widely as a function of factors such as stress and fatigue (Swain, 1982).

Whereas erroneous actions play a prominent role in applied disciplines such as human reliability analysis and cognitive engineering, the problem of modelling erroneous actions has rarely been adequately addressed by the HCI research community. Here investigators have tended to adopt general frameworks which portray the average user as someone who knows how to deal with information technology and who willingly participates in the interaction. Clearly, there are many situations where such a characterisation is appropriate. Increasingly, however, information technology is finding its way into application areas where the user is unfamiliar with it or may even be ill motivated.

The Accidental User

Discussions of the relation between artefacts and users often make a distinction between different user categories, such as novices and experts (Dreyfus & Dreyfus, 1980). This distinction refers to the degree of knowledge and skill that a user has *vis-à-vis* a specific artefact and makes the implicit assumption that the user is motivated to use the system. The spread of information technology, most recently as ubiquitous computing, however, means that there are many situations where users interact with information technology artefacts because they *have* to rather than because they *want* to. The possibility of doing something in another, more conventional – and presumably familiar – way has simply disappeared. Examples include finding a book in a library, paying bills, booking tickets and travels, etc. A trivial example is buying a train ticket since many stations no longer have a manned ticket booth but leave everything to machines. More complex examples are driving a train, flying an aircraft, or monitoring anaesthetics in the operating room (Woods et al., 1994). The number of cases will continue to grow because conventional modes of interaction disappear, often in the name of efficiency!

People who in this way are forced to interact with information technology may be called *accidental users*. An accidental user is not necessarily an infrequent or occasional user. In many cases the use may occur daily or weekly, but it is still accidental because there is no alternative. Neither is an accidental user necessarily a novice or an inexperienced user. For instance, most people are adept at getting money from an automated teller machine but the use is still accidental because the alternatives are rapidly disappearing. An accidental user is not necessarily unmotivated, although it may be the case. The motivation is, however, aimed at the product of the interaction rather than at the process of interacting *per se*. An accidental user can thus be defined as a person who is forced to use a specific artefact to achieve an end, but who would prefer to do it in a different way if an alternative existed. From the systemic point of view of the accidental user, the system is therefore a barrier that blocks access to the goal – or which at least makes it more difficult to reach the goal (e.g., Lewin, 1951).

The accidental user poses a particular challenge to the design of artefacts because most of the relevant disciplines, such as Human-Computer Interaction and Human-Machine Interaction, are predicated on the assumption that users are motivated and have the required level of knowledge and skills. In particular, models of 'human error' and human reliability implicitly assume that users are benign and only fail as anticipated by designers. A realistic approach to artefact design must challenge this assumption. One of the important applications of CSE is therefore how to foresee how the use of an artefact can go wrong.

User Models and Accidental Users

The notion of a user model looms large in all disciplines that deal with human-technology interaction. The purpose of a user model is to help the designer predict what the likely user reactions will be, hence to develop an interface and a dialogue flow that is optimal for the tasks. Newell (1993) has argued eloquently for the need to consider users that are temporarily or permanently impaired in their perceptual and motor functions. In addition to that, the notion of the accidental user argues that designers should consider how people will perform who have little or no motivation. In particular, designers should consider the following possibilities:

- That the user will misinterpret the system output, e.g. instructions, indicators and information. The misinterpretation need not be due to maliciousness, though that might sometimes be the case, but simply that users do not know what designers know, and that users do not see the world in the same way. A simple example is that the user has a different culture or native language and therefore is unfamiliar with symbols and linguistic expressions.
- That the user's response will not be one of the set of allowed events, hence not be recognisable by the system. System design can go some way towards preventing this by limiting the possibilities for interaction, but the variability of user responses may easily exceed the built-in flexibility of the system.
- That the user will respond inappropriately if the system behaves in an unexpected way. This means that the user may get stuck in an operation, break and/or restart a sequence any number of times, lose orientation in the task and respond to the wrong prompts, leave the system, use inappropriate modes of interaction (hitting it, for instance), etc.

In some sense, the accidental user should be considered as if governed by a version of Murphy's Law, such as: "Everything that can be done wrongly, will be done wrongly". More seriously, it is necessary to be able to model how actions are determined by the context rather than by internal information processing mechanisms, and to think of what happens when the context is partially or totally inappropriate. In order to design a system it is necessary to consider all the situations that can possibly occur. This means that the designer must be able to account for how users, whether accidental or not, understand and control the situation.

Traditional human factors models are reasonably effective when designers are interested in possible error modes specified in terms of their external manifestations. This is especially the case where the objective of analysis is to predict the probability of *simple* errors which may occur in relatively trivial tasks such as searching for a book in a library. Although more complex models of human behaviour have been specified at the level of observable behaviour, the basic limitation is that they provide little information regarding the psychological causes of erroneous actions. Such models are therefore unable to distinguish between an accidental and an intentional user. The analytic capability of human factors models is typically quite low and resultant analyses of failures in experimental systems tend to yield few general principles to help designers with the predicament of the accidental user.

Information processing models have a high analytic capability but are not very good at converting field data to useful and practical tools for prediction of possible erroneous actions. The analytic capability derives mainly from the large number of general statements relating to 'error' tendencies that are typically part of such models. The validity of predictions based on these models is, however, unclear. Experience shows that actions frequently fail when they should be well within the user's performance capability. Conversely, it is easy to document instances where user performance has been quite accurate for tasks where the model would predict failure (Neisser, 1982).

Models from CSE provide a better approach for characterising the interactions between information technology and accidental users – and intentional users as well. These models are particularly strong in their technical content because they are based on viable and well-articulated descriptions of human action – rather than of covert mental processes. Moreover, the emphasis on the contextual determination of human behaviour is clearly better suited to explanations of a user predominantly driven by environmental signals (e.g. the interface) as opposed to past experience and prior knowledge. The accidental user will typically have limited competence and limited control, and the ability of these models to describe how the interaction between competence, control and feedback determines performance is therefore essential. The CSE perspective also affords a comparatively high level of predictive capability in a model that can be converted into practical tools for investigation of erroneous actions. In our view it is therefore the all-round approach best suited to model how an accidental user interacts with information technology.

Chapter 6

Joint Cognitive Systems

Humans and artefacts have traditionally been considered as separate systems that perforce must interact. CSE proposes that the human-artefact ensemble be seen as a joint cognitive system in its own right, and that design and analysis starts from this level. Joint cognitive systems can be defined recursively and accentuate the necessity to provide clear definitions of the boundaries and the capabilities of the systems

INTRODUCTION

In Chapter 3, the concept of joint cognitive systems was introduced by referring to Stafford Beer's definition of a system, which emphasised the system's functions rather than its structure. For CSE, however, the most important thing is not that a system, such as the woman-*cum*-scissors, produces something but that it performs in a controlled or orderly manner. It is clear that control is required in order to produce something, although control by itself does not mean that something is produced – other than orderly behaviour. A line dancer, for instance, is in control of what s/he does, which is to keep balance and avoid falling to the ground, but there are no tangible products as a result of doing that.

According to the definitions given in Chapter 3, any JCS can be described on a lower level of aggregation in terms of the parts that together constitute the system, as well as a higher level of aggregation where the JCS in itself becomes part of a superordinate system. The decomposition or disaggregation ends with the simplest JCS, which consists either of two cognitive systems, such as two people working together, or of a cognitive system and an artefact, such as someone using a tool. The aggregation, on the other hand, has no natural upper limit but can go on as long as experience and imagination support it. Thus any JCS can be seen as constituting a part of a system on a higher level of aggregation. Whereas HMI usually has focused on the disaggregation, CSE considers aggregation and disaggregation to be equally important.

We can illustrate this by considering the situation of driving a car (Figure 6.1). Ergonomics and human factors have traditionally focused on the interaction between the driver and the car, such as how the instruments should look, how the controls should be designed, etc. A good example is the large number of studies that in recent years have looked at how competing interests for the driver's attention can affect driving (e.g., Alm & Nilsson, 2001). Whereas drivers previously could concentrate on driving and on staying within their field of safe driving (Gibson & Crook, 1936), they now have to cope both with a more complex traffic environment in terms of traffic density, speed of movement and various signs and signals, as well as with various forms of information and communication devices such as navigational support systems and mobile phones. (In addition to that, modern cars contain a number of sophisticated safety and support functions, which introduce a sneaking automation.) While it obviously is important that the ergonomics of the car, as the driver's workplace, is appropriate both from the classical anthropometric and the cognitive aspects, it is not sufficient. A car is designed to move or to serve as a means of transportation, and the driver-car system must therefore be seen within the context of driving.

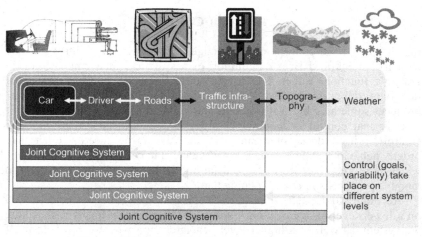

Figure 6.1: Joint cognitive systems for driving.

Relative to the driver-car system, the environment is – for argument's sake – made up of the streets and roads. But if the focus is changed from the level of the individual car-driver system to the cars moving on the roads, a new JCS comes into view. This is the car-driver-road system, which for convenience we may refer to as the traffic system. This JCS clearly has properties that are not found on the car-driver system level as can be seen from the issues of performance and control. For instance, effective performance is now expressed in terms of how well the traffic flows, which in turn is determined by the coagency among the individual car-driver systems, among other things. It therefore makes sense to study issues such as traffic flow and traffic structure, not just mathematically but also from a CSE perspective.

The traffic system, as a JCS, is defined within its environment, which we – again for the sake of the argument – assume to be the traffic infrastructure. This includes both tangible factors such as traffic signals, traffic signs and road markings, and intangible factors such as rules or laws of traffic, the Highway Code, etc. Just as it only made sense to study the car-driver JCS in the context of the traffic system environment, so it only makes sense to study the traffic system in the context of its environment, the traffic infrastructure, which we may call the overall transportation system. But the transportation system can also be described as a JCS in itself with functional characteristics that are different from those of the traffic system. For instance, the adequacy of road signs or the regulations governing working hours determine how effectively the transportation system functions. Yet even the transportation system must be seen relative to an environment, which we shall take to be the topography. Differences in topography may yield differences in performance of two transportation systems even if most of the features of the former are similar. The traffic rules, for instance, govern the traffic flow but not the topography.

The aggregation can be taken one step further by noting the specific characteristics of the car-driver-roads-infrastructure-topography system. Driving in mountains, for instance, is different from driving on a plateau. If the topography was beyond control, i.e., if it was a given that could not be changed in any way, then this aggregated larger system would not qualify as a JCS. But relative to the previous level of the transportation system, the topography can be changed, e.g., by building bridges or tunnels, by cutting through hills, etc. In that sense the topography is an artefact and in combination with the transportation system it constitutes a JCS on a higher level. That JCS clearly also exists in an environment, which we can take to be the weather. Here the aggregation comes to an end, for reasons that will be explained in the following.

On the Nature of System Boundaries

The distinction between a JCS and its environment is important in two different ways. First it is obviously important that a distinction is made, since without that it is unclear what the system really refers to. But it is also important to realise that this distinction is relative rather than absolute. In other words, the boundary is defined according to a set of criteria that depend on the purpose of the analysis hence on the system's function rather than its structure. In fact, the structural characteristics of a system depend to some extent on where the boundary is put. From a general point of view the distinction implies that the environment both provides the inputs to the system and reacts to outputs from the system. A more formal definition is as the following:

> For a given system, the environment is the set of all objects a change in whose attributes affect the system and also those objects whose attributes are changed by the behavior of the system. (Hall & Fagen, 1968, p. 83)

It may rightfully be asked why objects that react to the system, as described by the above definition, are considered as the system's environment instead of simply being included in the system. Yet to do so would only move the boundary further outwards without solving the demarcation problem. It may also rightfully be asked whether the environment itself has a boundary, or whether it is infinitely large. Answers to these questions can best be found by taking a pragmatic rather than a formal stance. Indeed, the boundary of a system cannot be determined in an absolute sense but relatively only to a specific purpose or aim. The determination is in practice made by limiting the system to those objects that are important for its functioning and excluding those that are not, all relative to the purpose of the analysis. Within CSE the definition can be improved by considering whether the 'objects' referred to by Hall & Fagen (*op. cit.*) are something that the JCS can control. If so, they should be seen as part of the system; if not, they should be seen as part of the environment.

The boundary of a JCS can be defined pragmatically by considering two aspects. The first is whether the functioning of an object is important for the JCS, i.e., whether the object constitutes a significant source of variability for the JCS. The second is whether it is possible for the JCS effectively to control the performance of an object so that the variability of the object's output remains within a pre-defined range. By using these criteria, we may propose a pragmatic definition of how to determine the boundary of a JCS, as indicated in Table 6.1. More loosely, the boundary is defined by the fact that it makes sense, *vis-à-vis* the purpose or goals of the JCS, to consider the JCS as a unit. The fact that the definition is relative does not necessarily make it imprecise.

The definition contained in Table 6.1 also makes it clear why in the driving example the aggregation that started with the car-driver system had to stop at the level of the weather. The weather belongs to the category of objects that are important for the JCS's ability to maintain control, yet which cannot effectively be controlled by the JCS.

AUTOMATION AND JOINT COGNITIVE SYSTEMS

While it always is important to be precise in determining the system boundaries, it is vital in relation to automation. Automation is a central topic in human-machine systems research in general and in CSE in particular. It is also a topic about which much has been written over the years, both works that provide an overall perspective (Billings, 1996; Dekker & Hollnagel, 1999; Parasuraman & Mouloua, 1996; Sheridan, 1992 & 2002) and works that go into the details (Bainbridge, 1983; Moray et al., 1995; Sarter & Woods, 1997). For the purpose of the present discussion, we shall focus on how the various design principles for automation – often referred to as automation philosophies – relate to the concept of JCSs.

Table 6.1: A Pragmatic Definition of the JCS Boundary

	Objects whose functions are important for the ability of the JCS to maintain control.	Objects whose functions are of no consequence for the ability of the JCS to maintain control.
Objects that can be effectively controlled by the JCS.	1. Objects are included in the JCS	2. Objects may be included in the JCS
Objects that cannot be effectively controlled by the JCS	3. Objects are not included in the JCS	4. Objects are excluded from the description as a whole.

Automation can be defined as the execution by a mechanism or a machine of a function that was previously carried out by a human (Parasuraman & Riley, 1997). Although we have come to see automation as intrinsically linked to the use of computers and computing technology, increasingly in the form of vanishingly small and specialised components, the history of automation goes back to the water clock invented by Ktsebios in the third century BC. This was more than 2.000 years before the advent of computers – even if we take Joseph Jacquard's loom controlled by punched cards (1801) as an early example of the latter.

From its very beginning automation depended on the use of mechanical devices to work – and, of course, on the ability of humans to think of clever ways of creating those devices. This dependency continued throughout the industrial revolution and well into the 20th century. In practice this meant that automation could be applied only in industrial contexts, where the cost of building the automation was small relative to the cost of the production facility, and where it could easily be fitted into a place of work. (Exceptions were exclusive toys built for the pleasure of royalty and the very rich, e.g., Wood, 2002. Even today such toys remain rather expensive, e.g., the Sony Aibo). Since the speed of industrial processes remained fairly low, there were no automation-induced problems; i.e., operators were able to keep pace with the speed of machines and automation was rarely so complex that operators had problems in understanding how it worked – or in anticipating what it would do.

All this changed around the middle of the 20th century, which saw the development of the first digital computer (the ENIAC) in 1945, the creation of Cybernetics, an exhaustive mathematical analysis of the theory of feedback and control that provided the formal basis for self-regulation and automation (Wiener, 1965; org. 1948), and the formulation of the mathematical theory of communication (Shannon & Weaver, 1969; org. 1949) – commonly known as information theory. In combination with two major technological inventions – the transistor by William Shockley in 1947 and the integrated circuit by Jack Kilby in 1958 – this meant that automation could be made far more complex, that it became much faster, that less power was needed for it to work, and that it took up less space. The limits to the capability of automation were now set by the designers' imagination rather than by physical constraints or production costs. Altogether this increased the pace by which processes could be run. In parallel to that the ability to send both sensor inputs and control outputs over very long distances meant that it suddenly became possible to extend automation both in breadth and in depth. The horizontal expansion meant that a large number of subsystems could be put under common control, while the vertical expansion meant that each subsystem or process could be automated on several levels or layers – from minute process details to scheduling and production planning. Since the cost of automation at the same fell dramatically, it could suddenly be applied everywhere. This has changed the rules of the game as far as human-technology interaction is concerned, and therefore also of how automation design should be approached.

From the industrial revolution and onwards, automation was seen from the perspective of the process or function to be controlled, with little or no concern for the people involved. This was a perfectly reasonable thing to do, both because machines still were somewhat primitive and because the speed as well as complexity of automation was well within what operators could handle. This led to an approach that became firmly entrenched and which therefore became the 'natural' way of designing automation. It was therefore continued even when the nature of automation changed. Another reason for the dominance of a technological perspective was that automation from the very beginning was a problem for engineering rather than for behavioural science. This attitude did not change until well into the 1950s, when human factors engineering came onto the scene – partly because of problems that followed the developments mentioned above.

The situation today is that automation is omnipresent and that it continues to expand both horizontally and vertically. Yet the approach to design of automation at work in production, transportation, and communication has changed less rapidly than the technology – if it has changed at all! This has led to a growing number of problems which show themselves as day-to-day disturbances and which sometimes lead to dramatic accidents. The response to such events has, ironically, often been to introduce even more automation to reduce the dependence on what is seen as unreliable human performance (e.g., Bainbridge, 1983; Hollnagel, 1999b, cf. also the description of the self-reinforcing complexity cycle in Chapter 1).

Degrees of Human-Machine Dependency

The development of automation, particularly since the 1950s, has been a major influence on the relations between humans and machines, and therefore also on how we view human-machine systems. The philosophical side on this issue has already been mentioned in Chapter 2, in the discussion of the use of artefacts. A more detailed and practical description, proposed by Sheridan (2002), provides a characterisation of the relationship between humans and automation, with the following stages (cf. Figure 6.2).

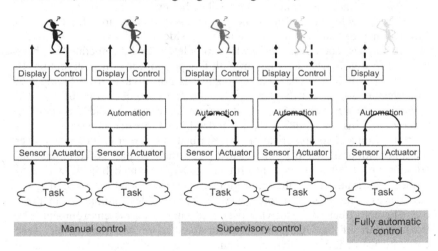

Figure 6.2: Stages of human-machine dependency.

In the first stage of development, in Figure 6.2 called manual control, humans were in direct control of the process. Sensor data (i.e., measurements) were presented directly to the operator and the resulting actions similarly worked directly on the process. The JCS was therefore the human (operator) and the technology, specifically information presentation devices and control devices. Although simple automation was used from the very beginning both to help regulate the process, e.g., James Watt's flying-ball governor, and to monitor when sensor data exceeded predefined limits, e.g., as in alarm systems, this only affected basic sensory-motor functions that operators were bad at doing anyway, and did not really impinge on their ability to control.

In the next stage, in Figure 6.2 called supervisory control, automation progressively took over functions that hitherto had been carried out by people, and thereby gradually removed them from direct control of the process. This is shown in Figure 6.2 both by a coupling from sensors to actuators within the automation itself and by a reduction of the interaction between process and operator (shown by dotted lines). In both cases these developments started modestly but rapidly grew as the technology matured. In practice this illustrates a condition where the automation by itself manages an increasing number of events and functions, at least as long as the system remains in a normal state. At this stage of development the nature of the JCS also changed. It became more complex partly because automation in many cases functioned as a quasi-cognitive artefact that could control a process, although it could do it only in an inflexible manner, and partly because the boundaries of the JCS changed, as described below.

As the development of technology continued, this led to a final stage of full automation shown in the right-hand side of Figure 6.2. Here the automation has taken over more or less completely, and the operator has been reduced to follow what is going on with few possibilities for intervention. In this case the JCS also changed, although it is not obvious from Figure 6.2. The change was primarily due to the fact that the boundaries of the JCS expanded. At the same time as automation increased, so did the coupling among systems or units. This is, indeed, one of the consequences of centralisation and computerisation as described in Chapter 2. Since the power of automation inevitably is limited, cf. the 'ironies of automation' (Bainbridge, 1983), one way to protect against unwanted breakdowns is to reduce the variability of the environment. This was done by expanding the boundaries of the JCS, in accordance with the principles summarised in Table 6.1. The net effect was that what previously had been the environment now became part of the JCS, thus leading to a more complex whole. One consequence of this was that the meaningful level of analysis had to change to a more aggregated level.

While the problems of overt automation in industry has got the lion's share of attention in the research community – as well as in the public eye – it should not be ignored that covert or sneaking automation relentlessly is entering daily life in a number of ways. Examples are found in service functions, transportation, and communication, and increasingly also in homes and at leisure. A good illustration is provided by a modern automobile – which quite appropriately can be described as a computer network on wheels.

Modern cars comprise a considerable number of automated functions ranging from anti-lock braking systems (ABS) and electronic stability programmes (ESP) to adaptive cruise controls (ACC) and lane departure warning systems (LDWS). All of these are activated automatically, usually without prior warning and sometimes to the surprise of the driver (Broughton & Baughan, 2002). In the case of ABS and ESP this is clearly necessary, since the time needed to respond is in the order of a few hundred milliseconds, hence too fast for most people. In case of the warning systems, such as ACC and LDWS, the driver is supposed to take action, but should s/he fail to do so the safety systems may themselves take over control of the car. All this is done to enhance safety, which is a noble purpose indeed. Yet little thought has gone into how the driver and the automation can successfully co-exist, not least when several systems become active at the same time, as well they may. Even worse, the semi-autonomous behaviour of individual car-driver systems may have serious consequences for the traffic system. Knowing how to deal with this is therefore an important objective for CSE.

Automation Philosophies

As argued above, automation design was from the start an engineering concern and this established a tradition for automation design that never has been seriously challenged. The fundamental shortcoming of this tradition is the unchallenged assumption that humans and machines should be described as separate but interacting systems. In contrast to that CSE proposes that humans and machines should be seen as constituting a JCS, and that the emphasis accordingly should be on coagency rather than on interaction.

The Left-Over Principle

Discussions of automation have been central in human factors engineering since the late 1940s, usually under the topics of function allocation or automation strategies. Over the years several different principles for automation design have been suggested although they all – explicitly or implicitly – have embraced the above-mentioned assumption about the nature of the human-machine relation. According to the simplest approach, commonly known as the *left-over* principle, the technological parts of a system should be designed to do as much as feasible (usually from an efficiency point of view) while the rest should be left for the operators to do.

The appealing 'logic' behind this approach is that machines are capable of doing a great many things faster and more reliably than people can. In combination with the ever increasing demand for efficiency and the rising cost of human labour this leads engineers to try to automate every function that can be. No less a person than Chapanis (1970) defended this approach and argued that it was reasonable to 'mechanize everything that can be mechanized.' The proviso of this argument is, however, that we should mechanise everything that *can* be mechanised, only in the sense that it can be guaranteed that the automation or mechanisation always will work correctly and not suddenly require operator intervention or support. Full automation should therefore be attempted only when it is possible to anticipate every possible condition and contingency. Such cases are unfortunately few and far between, and the available empirical evidence may lead to doubts whether they exist at all.

Without the proviso, the *left-over* principle implies a rather cavalier view of humans since it fails to include any explicit assumptions about their capabilities or limitations – other than the sanguine hope that the humans in the system are capable of doing what must be done. Implicitly this means that humans are treated as extremely flexible and powerful machines, which at any time far surpass what technological artefacts can do.

Since the determination of what is left over reflects what technology *cannot* do rather than what people *can* do, the inevitable result is that humans are faced with two sets of tasks. One set comprises tasks that are either too infrequent or too expensive to automate. This will often include trivial tasks such as loading material onto a conveyor belt, sweeping the floor, or assembling products in small batches, i.e., tasks where the cost of automation is higher than the benefit. The other set comprises tasks that are too complex, too rare or too irregular to automate. This may include tasks that designers are unable to analyse or even imagine. Needless to say that may easily leave the human operator in an unenviable position.

The Compensatory Principle

A second approach to automation is known as the *compensatory* principle or Fitts' List (after Fitts, 1951). In this approach the capabilities (and limitations) of people and machines are compared on a number of salient dimensions, and function allocation is made to ensure that the respective capabilities are used optimally. The original proposal considered the following dimensions: speed, power output, consistency, information capacity (transmission), memory, reasoning/computation, sensing, and perceiving. Since function allocation is based on the juxtaposition of human and machine capabilities, the approach is sometimes ironically referred to as the 'Men-Are-Better-At, Machines-Are-Better-At' or 'MABA-MABA' strategy (Dekker & Woods, 1999).

The compensatory approach requires that the situation characteristics can be described adequately *a priori*, and that the variability of human (and technological) capabilities will be minimal and perfectly predictable. Human actions are furthermore seen as mainly reactive to external events, i.e., as the result of processing input information using whatever knowledge the operator may have – normally described as the mental model. It also implies that it makes sense to decompose the interaction and the capabilities of humans and machines into smaller elements. The determination of which functions to assign to humans and which to machines is, however, not as simple as the compensatory principle and the categories of Fitts' list imply. Rather than considering functions one by one, the nature of the situation, the complexity, and the demands must also be taken into account. Furthermore, the principle of compensation disregards the higher order need for function co-ordination that is required for a system to achieve its goals. There are, of course, some low-level, sensory-motor functions where a substitution is both possible and desirable, since treating humans as very simple mechanisms is both demeaning and unproductive. Yet function allocation cannot be achieved simply by substituting human functions by technology, nor *vice versa*, because of fundamental differences between how humans and machines function and because functions depend on each other in ways that are more complex than a mechanical decomposition can account for. Since humans and machines do not merely interact but act together, automation design should be based on principles of coagency. The issue is one of overall system design and the functioning of the JCS, rather than the allocation of functions as if humans and machines were independent entities.

One expression of that is the demand-capacity match shown in Figure 6.3. The design of workplaces in general and automation in particular is often done to alleviate the demand-capacity gap discussed in Chapter 1. When this is done with a view to process efficiency, i.e., on the premises of technology rather than the premises of humans, the result is often that tasks are designed so that the minimum demands require maximum capacity. This means that humans are supplemented – or replaced – by automation to ensure that the task can be accomplished under optimal conditions of work. What should rather be done is to design tasks so that the maximum demands can be met by the normal capacity – or preferably even by the minimum capacity. The reason for advocating this approach is that the demands can vary considerably, and that situations inevitably will occur where the normal capacity is insufficient. Quite apart from that, demands will invariably increase as described by the Law of Stretched Systems described in Chapter 1.

The Complementarity Principle and Function Congruence

A third approach, which has emerged during the 1990s, is called the *complementarity* principle (Grote et al., 1995; Wäfler et al., 2003). This approach aims to sustain and strengthen the human ability to perform efficiently and therefore considers the work system in the long term, including how work routines change as a consequence of learning and familiarisation. The main concern is not the temporary level of efficiency (or safety), but rather the ability of the system to sustain acceptable performance under a variety of conditions. The effects of automation are by no means confined to the moment, but may reach far into the future, cf. the substitution myth described in Chapter 5. The effects may be transitory, such as variations in workload or task complexity; they may be short-term, such as effects on situation awareness or fatigue (Muscio, 1921); and they may be in the medium- to long-term such as trust, self-confidence and the level of skills. While automation design mostly has considered the transitory or short-term effects, it is clearly also necessary to consider the consequences over a longer time span.

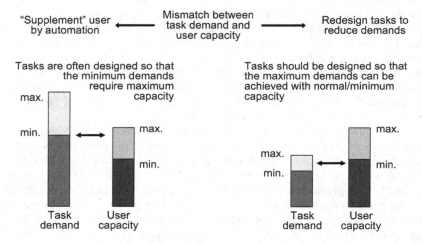

Figure 6.3: Two types of demand-capacity match.

The complementarity principle is a move in the right direction because it looks at human-machine cooperation rather than human-machine interaction and because it emphasises the conditions provided by the overall socio-technical system. It is thus consistent with the view of CSE, although in its development it represents a different perspective. It is, however, easy to translate the complementarity principle into the terminology of CSE and JCS. This can be expressed as an approach called *function congruence* or *function matching* (Hollnagel, 1999b). Function congruence is based on recognising the dynamics of work, specifically the fact that capabilities and needs may vary over time and depend on the situation. One way of offsetting that is to ensure an overlap between functions as they are distributed within the JCS, corresponding to having a redundancy in the system. This provides the ability to redistribute functions according to need, hence to choose from a set of possible function allocations. The net desired effect is an improvement in the overall ability to maintain control.

Summary

The common basis for introducing automation is still to consider the operator's activities in detail and evaluate them individually with regard to whether they can be performed better (which usually means more cheaply) or more safely by machines or technology. An example of that are the attempts to monitor workload and attention during driving, with the implication that the car in some way can intervene if the driver is deemed unable to respond correctly. Although some kind of decomposition is necessary, the disadvantage of considering functions one by one – in terms of how they are accomplished rather than in terms of what they achieve – is the inevitable loss of the view of the human-machine system as a JCS. The level of decomposition is furthermore arbitrary, defined, e.g., by the granularity of models of human information processing or by classifications based on technological analogies as epitomised by the Fitts list. Doing so fails to recognise that human abilities generally are flexible and adaptive, hence of a heuristic nature, rather than steady and stable hence of an algorithmic nature. This means that they are difficult to formalise and implement in a machine. Admittedly, the use of heuristics may also sometimes lead to unwanted and unexpected results but probably less often than algorithms do. The use of heuristics is nevertheless the reason why humans are so efficient and why most systems work.

The characteristics of the three automation philosophies presented above are summarised in Table 6.2. In relation to the notion of JCSs, only the third philosophy begins to address these issues. In addition to the categories listed in the left-hand column of Table 6.2, it is also relevant to consider the reasons for introducing automation. In the case of *evolutionary design*, improvements are introduced when they become available and affordable. As described by the self-reinforcing complexity in Chapter 1, automation is driven by technological innovation – and often by promised advantages rather than established findings. In the case of *reactive design*, automation is driven by need to avoid past failures. Here risks are removed after accidents by automating manual functions, on the assumption that 'human error' was the root cause. Despite the obvious simplicity of this idea, it has in several cases been officially espoused as a design principle, and can still be found as an unspoken assumption of many accident analysis methods (e.g., Hollnagel, 2004).

Ironies of Automation

The ironies of automation have been mentioned several times already. The concept refers to a seminal paper in the human factors literature (Bainbridge, 1983), which provided a lucid analysis of the background and principles of industrial automation, in particular the view that operators are unreliable and inefficient and therefore should be eliminated from the system. The ironies of this attitude are twofold. The first irony is that design decisions themselves can be a major source of operating problems, for instance because they are based on oversimplified assumptions about human capabilities or the nature of the interaction. The second irony is that any attempt to eliminate the operator still leaves the operator to carry out the tasks that designers are unable to automate. It was noted above that full automation, i.e., to 'mechanize everything that can be mechanized', is a valid principle only if it can be assured that nothing is missed. This is, however, a virtual impossibility, and the consequence is that operators always will have to face situations that have defied analysis. This is, in fact, not only a problem for automation design but also for display design, as noted in Chapter 4.

Table 6.2: Summary of Automation Philosophies

	Leftover principle:	Compensatory principle (Fitts' list):	Complementarity – congruence
Theoretical perspective	Classical human factors	Human machine interaction	Cognitive systems engineering
Function allocation principle	Leave to humans what cannot be done by technology	Avoid excessive demands to human performance	Enhance coagency, support long-term comprehension
Purpose of function allocation	Ensure process efficiency by automating whatever is feasible	Ensure efficiency of human-machine interaction	Enable JCS to maintain control under different conditions
View of operator (model)	None	Limited capacity IPS	Cognitive system
System description	Independent entities	HMS	JCS
Assumptions about interaction	Independent entities	Valid *a priori* description	Coagency

From a historical perspective, automation has been used either to ensure a more precise performance of a given function or to improve the stability of system performance. After the industrial revolution, a further motivation was to increase the speed of work and production. The net effect of the increasing use of technology was that humans gradually came to be seen as bottlenecks for system performance, thereby strengthening the need of further automation. It is no coincidence that human factors engineering emerged in the late 1940s, when the rapid changes brought about by the scientific and technological developments in the preceding decade had undermined the hitherto peaceful co-existence between humans and machines. These developments opened the field for a new type of engineers branded rather appropriately by Norbert Wiener as "gadget worshippers" (Wiener, 1964, p. 53), who "regard(ed) with impatience the limitations of mankind, and in particular the limitation consisting in man's undependability and unpredictability" (*ibidem*). Humans were not only a bottleneck for system performance but also constituted a source of unwanted variability that might be the cause of incidents and accidents. According to the logic of this line of reasoning, it made sense to replace humans by automation as far as possible. Unfortunately, this created many of the problems we face today.

Lessons of Automation

The use of automation invariably confronts two views or philosophies. One view refers to the fact that the number of events with unwanted consequences is steadily rising. In these cases the human is often seen as a cause of failures and as a limiting factor for system performance. The solution is therefore to reduce the role of the operator by increasing the level of automation, or preferably to eliminate the need for the operator altogether. Although this in principle may seem to be an appropriate arrangement, there are some sobering lessons to be learned from the study of the effects of automation (e.g. Wiener, 1988). These are, briefly put, that:

- Workload is not reduced by automation, but only changed or shifted.
- Human 'errors' are not eliminated, but their nature may change. Furthermore, the elimination of small erroneous actions usually creates opportunities for larger and more critical ones.
- There are wide differences of opinion about the usefulness of automation (e.g., in terms of benefit *versus* risk). This leads to wide differences in the patterns of utilisation, impinging on the actual effects of automation.

Another view acknowledges that it is impossible to consider every possible contingency that may arise during the use of a system, and that it therefore is impossible for automation completely to take over human functions. Humans are furthermore seen as a source of knowledge, innovation and adaptation, rather than just as a limiting factor. The conclusion from this perspective is that automation should be made more effective by improving coagency, the coupling between man and machine, a JCS view. Two main issues in the practical implementation of this view are: (1) what humans should do relative to what machines should do, and (2) whether humans or machines are in charge of the situation. The first refers to the function allocation proper, and the second to the issue of responsibility. The two issues are obviously linked, since any allocation of functions implies a distribution of the responsibility. The responsibility issue quickly becomes complicated when function allocation changes over time, either because adaptation has been used as a deliberate feature of the system, or – more often – because the system design is incomplete or ambiguous. Although both issues have been most conspicuous in relation to the use of automation, the problem of function allocation is fundamental to any kind of human-machine interaction, from display design to adaptive interfaces.

One attempt to resolve the automation design dilemma, for such it must be called, is to resort to human-centred automation. Yet this often raises more problems than it solves, mainly because human-centredness is an attractive but ill-defined concept. There are three typical interpretations:

- That automation design is based on human needs. This is known rarely to be the case; design is usually due to technical needs.
- That people 'in the loop' are part of the development of the design (= user participation in design).
- That design is based on predicted improvements to human cognition and performance. This requires well-developed models and an understanding of the nature of work.

The CSE alternative to human-centred design is that designers from the beginning consider the impact of changes on the roles of people in the system and on the structure of the larger socio-technical system. In other words, they should confront the substitution myth and challenge it. This can be done by adopting a JCS perspective and by describing the system in terms of goals and the functions needed to achieve them, rather than in terms of components and their capabilities. In other words, by adopting a top-down rather than a bottom-up approach. In this way fixed descriptions of task sequences are replaced by definitions of the functions that are needed at each point in time. The JCS should at every point in time have sufficient capabilities to achieve the goals while being able to anticipate and predict future events. The best possible candidate for a top-down analysis principle is the goals-means method, which will be described in the following.

FUNCTION ANALYSIS AND GOAL ACHIEVEMENT

The design of any human-machine system, and in particular the design of complex systems, requires some way of determining how tasks and functions shall be distributed or assigned. This in turn requires some way of identifying which tasks and functions are needed for the system to fulfil its functions, specifically to achieve its stated goals. A task or function allocation always assumes an *a priori* description of tasks or functions, that can be assigned to either humans or machines. In the case of a JCS, the need is to define the functions and capabilities that are necessary for the JCS to maintain control of what it does.

The normal way of approaching this is by means of a task analysis, but a task analysis logically requires as a starting point a description of the system as already decomposed into humans and machines with their component functions. Whereas a task description can be useful for improvements (as well as for many other purposes), system design requires a functional analysis rather than a task analysis. Hierarchical task analysis exemplifies a decomposed view, with functions allocated to humans and machines (Annett, 2003). Other approaches, such as CWA (Vicente, 1999) looks at demands rather than existing tasks, but does so from the view of a hierarchical set of concepts, which means that it becomes difficult to account for variable couplings and dependencies among functions.

Goals and Means

The alternative to a hierarchical analysis is to use a goals-means analysis. The goals-means principle was used as the main strategy in the General Problem Solver (GPS; Newell & Simon, 1961), although here it was called the means-ends analysis. (In the two expressions, 'goals' and 'ends' are basically synonymous, but the order between goals/ends and means has been reversed.) The basic principle of the means-ends analysis was that differences between the current state and the goal state were used to propose operators (meaning symbols that represent operations, rather than humans that work), which would reduce the differences. In the GPS, the correspondence between operators and differences was provided as knowledge in the system (known as a Table of Connections). The means-ends principle was later used as the basis for the GOMS method (Card et al., 1983).

The General Problem Solver was by no means the first use of goals-means analysis in behavioural science. In the 1950s several psychologists had become interested in the developments in the meta-technical sciences, in what became the forerunner of cognitive psychology and cognitive science. Many researchers at that time felt that cybernetics could be a useful basis for describing human behaviour, as demonstrated by Boring's (1946) deliberations about mind and mechanism. A different approach was that taken by Miller, Galanter and Pribram who in their influential book on *Plans and the Structure of Behavior* declared their fundamental interest to be:

> ... to discover whether the cybernetic ideas have any relevance for psychology ... There must be some way to phrase the new ideas so that they can contribute to and profit from the science of behaviour that psychologists have created. (Miller, Galanter & Pribram, 1960).

Simple Test-Operate-Test-Exit (TOTE)

Their proposal for how to phrase the new ideas was the Test-Operate-Test-Exit (TOTE) framework. TOTE demonstrated how a goals-means decomposition could be used as the basis of a functional model of behaviour, and thus preceded Newell & Simon's use of it for the structural analysis of reasoning. The work on plans by Miller et al. clearly showed that cybernetic ideas and concepts could be used by psychology and was one of the more notable influences for the emerging cognitive psychology.

The example that Miller et al. used was the simple action of hammering a nail into a piece of wood. This was described in terms of a plan for hammering, which is shown graphically in Figure 6.4. The goal is that the nail is flush with the surface of the piece of wood, while the means is to hammer down the nail. The plan (or procedure) is to continue hammering until the goal condition has been achieved. In this case the plan has the following components: (1) test if the head of the nail is flush (Test), (2) if not then hit the nail with the hammer (Operate), (3) if so then stop hammering (Exit); otherwise repeat the plan.

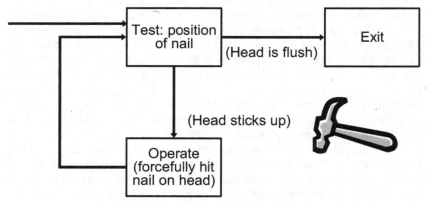

Figure 6.4: TOTE for hammering a nail.

One of the more powerful features of this description is that the activity, i.e., hammering, is the result of a plan or an intention, rather than the response to a stimulus. The person obviously can see that the head of the nail sticks up, but there is no simple way of explaining how this 'stimulus' leads to the response of hitting the nail with the hammer. From a contemporary point of view the description of TOTE is simple and obvious and may seem to be no more than a basic conditional statement. Yet it is important to consider what the level of common sense knowledge was when it was developed in the late 1950s. At that time few people had any experience with programming or flow chart descriptions, or even of thinking in terms of algorithms.

The value of TOTE is, of course, not just that it can be used to describe how a nail can be hammered into a piece of wood, but that it provides a general framework for describing any kind of activity or behaviour. The TOTE unit is functionally homomorphic to the feedback control loop, which in the 1960s was commonly known among control engineers and neurophysiologists, but not among psychologists. TOTE represents a recursive principle of description, which can be applied to analyse an activity in further detail. (That Aristotle's description of reasoning about means, mentioned in Chapter 3, was also recursive shows how fundamental this principle is.)

The goals-means decomposition for TOTE can, of course, be continued as long as needed. In the case of plans there is, however, a practical stop rule. Once a means describes an action that can be considered as elementary for the system in question, the decomposition need go no further. As simple as this rule may seem, it has important consequences for system design in general, such as the expectations to what the user of an artefact is able to do, the amount of training and instruction that is required, the 'intuitiveness' of the interface, etc. The advantage of a recursive analysis principle is that it forces designers to consider aspects such as stop rules explicitly; in hierarchical analyses the stop rule is implied by the depth of the hierarchy and therefore easily escapes attention.

Recursive Goals-Means Descriptions

A more complex version of the goals-means principle makes a distinction between three aspects called goals, functions (or activities), and particulars (or mechanisms). In the language of functional analysis this corresponds to asking the questions 'why', 'what', and 'how', as illustrated by Figure 6.5. The goals provide the answer to the question of 'why', i.e., the purpose of the system. The functions describe the activity as it appears and provides the answer to the question of 'what', i.e., the observable behaviour of the system. Finally, the particulars provide the answer to the 'how', i.e., by giving a detailed description of the way in which the functions are achieved. Figure 6.5 also illustrates how the goals-means decomposition can be used recursively. That which constitutes the means at one level, becomes the goals at the next level down, and so on. Altogether this provides a method for functional decomposition, which is essential to CSE because it logically precedes the structural decomposition that is commonly used in systems analysis. The goals-means analysis has been used extensively as the basis for the engineering analysis method known as multi-level flow modelling (MFM; cf. Lind & Larsen, 1995), for instance as a basis for constructing correct procedures. For the purpose of illustrating the goals-means analysis in CSE, it is nevertheless sufficient to stay with the basic goals-means decomposition method, since the means always can be decomposed further into functions and particulars.

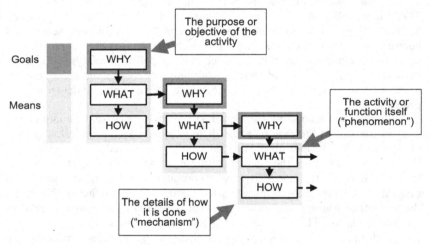

Figure 6.5: The goals-means decomposition.

The advantage of a recursive analysis principle is that the resulting description is as simple as possible with regard to the concepts and relations that are used. In contrast, an analysis based on a hierarchical principle must refer to a pre-defined structure of a finite, but possibly large, set of concepts organised at a fixed number of levels. Each level typically represents a different mode of description, which means that the descriptive dimensions can be radically different at the top and the bottom (Lind, 2003). A hierarchical analysis furthermore has a pre-determined depth, corresponding to the number of levels in the hierarchy, and usually tries to achieve the maximum depth. Since a recursive analysis principle does not imply a pre-defined depth or number of steps, it compels the analyst to be parsimonious with the number of categories, and makes a virtue out of using as few categories as possible. In fact, by focusing on the functional rather than the structural relations between the levels of the analysis, it is possible to avoid characterisations of the levels in absolute terms.

The disadvantage of a recursive principle is that there is no obvious minimum or maximum depth and therefore no fixed number of levels. In order for this to be practically manageable, a clear stop rule must be defined for each type of application. For the purpose of task analysis, a stop rule could be that the analysis reaches what corresponds to the level of basic operator competence. For the purpose of accident analysis, a stop rule could be that there is insufficient information to make a justifiable choice or inference. This stop rule has been used in the retrospective version of the Cognitive Reliability and Error Analysis Method (Hollnagel, 1998a).

Chapter 7

Control and Cognition

The main thrust of CSE is how control can be improved. This chapter discusses the essential issues in control, such as closed- and open-loop control, and how automation and technological support fit into that. CSE applies a basic cyclical model, which can be extended to account for multiple simultaneous layers of control. This is necessary to give a realistic description of how a joint cognitive system functions.

INTRODUCTION

The concept of control is essential to CSE since it is part of the definition of what a cognitive system is (cf. Chapter 1). Control is one of those everyday concepts that are meaningful to all and we have so far used the term control without formally defining it. Before going into a more detailed description of how control is modelled in CSE, it is therefore appropriate to provide a definition of the term.

One basis for that can be found in *The Mathematical Theory of Communication* (Shannon & Weaver, 1969; org. 1949), which includes a description of communication problems on three different levels. On Level A, the technical problems are about how accurately the symbols of communication can be transmitted. On Level B, the semantic problems are about how precisely the transmitted symbols convey the desired meaning. Finally, on Level C, the pragmatic problems are about how effectively the received meaning affects conduct in the desired way. This suggests that the essence of control is the ability to affect the conduct of the recipient in the desired way and thereby achieve a desired effect. Communication theory refers to a sender and a receiver, both of which are assumed to be human beings. CSE uses a broader definition where the sender can be a cognitive system or a JCS, while the receiver can be a cognitive system, a JCS, or an artefact. Table 7.1 provides some examples of control from a CSE perspective, where the left-hand column shows the controlling system (sender) and the upper row the target (receiver):

Table 7.1: Examples of Control in a CSE Context

	Cognitive system	Joint cognitive system	Technological artefact
Cognitive system	Person-to-person communication: giving an order; trying to elicit a favour, ...	Commanding a unit, running an enterprise, getting something done, ...	Getting something to work, setting up a system, ...
Joint cognitive system	Instructions, rules, guidelines, procedures, ...	Running a public service, such as a train system, maintaining an aeroplane, ...	

Simply speaking, the problem of control is how to obtain a given outcome either from a technical artefact, such as a machine or device, from another person, or from an organisation. Looked at in this way, most of what we do in daily life can be described as an issue of control, understood as the ability to produce or achieve a desired outcome. This can be done by the cognitive system itself when it has full control of the situation or may require the collaboration and cooperation with other cognitive systems, in which case it can be said to have partial control of the situation.

To be in control is, however, more than being able to achieve intended outcomes, i.e., to get what you want. The definition of a cognitive system included the term anti-entropic, in the meaning of bringing into order. To be in control therefore requires that the system is able not just to bring about desired outcomes but also to prevent or recover from anti-entropic events, i.e., to cancel or neutralise possible disturbances and disruptive influences. For the purposes of CSE, control can be defined as the ability to direct and manage the development of events, and especially to compensate for disturbances and disruptions in a timely and effective manner. This means the ability to *prevent* unexpected conditions from occurring, as well as effectively to *recover* from them should they occur.

Feedback and Control

Control requires the ability to compensate for differences between actual and intended states. This in turn requires the ability somehow to sense, measure, or perceive the difference. In order for this to be possible, the difference must first of all be above a certain threshold value. Secondly, the JCS must then notice this and, thirdly, interpret it correctly. While this may seem a trivial matter if we simply think of measurements exceeding a threshold, it is far from trivial when it comes to human behaviour, since humans have a wonderful – and many times also useful – ability to disregard facts. From a philosophical perspective, it is the question of when a person realises that a situation or condition is incorrect. Since cognitive systems actively seek rather than passively receive information, they can clearly sometimes seek the 'wrong' information and thereby end up going in the wrong direction.

The failure to recognise that a difference exists is best known from cases of information input overload (cf. Chapter 4). One of the costs – or risks – of applying any of the associated coping strategies is that information is lost either because it is omitted, because the precision has been reduced, because of filtering or cutting categories, or because of queuing. The benefit is, of course, that enough information is retained to enable the system to continue as planned.

Failing to recognise a difference may even occur when conditions are normal, as vividly illustrated by change blindness. This is a phenomenon in visual perception in which very large changes, occurring in full view in a scene, are not noticed. It started out as a curious experimental finding (McConkie & Currie, 1996) but soon became a widely recognised phenomenon that challenged the conventional information processing theories of perception (e.g., O'Regan & Nöe, 2001). It also quickly found its way into practical studies, for instance, as a way of explaining lapses of attention during driving (e.g., Batchelder et al., 2003) – or the more general 'looked but did not see' phenomenon.

Classical information processing theories have described the human as responding to incoming information, possibly with some low level filtering mechanisms to make a pre-selection of what the person becomes conscious of a moment later (Broadbent, 1958; Moray, 1970). Present day information processing theories have adopted this tradition in the sense that humans are modelled as responding to the information presented rather than as selecting it. CSE has adopted the more active view according to which humans actively search for and select information. In other words, perception is active and guided rather than passive and responsive. One example of this is the way in which people narrow their search to include only what they expect, which in terms of control means focusing on anticipated responses. There are obvious advantages to such a strategy, but also obvious disadvantages. In the CSE cyclical model it can be illustrated as shown in Figure 7.1.

Feedforward and Control

While feedback control is necessary, it is not efficient if the time required to analyse the feedback is so long that new information arrives before the analysis has finished, or if the system changes so rapidly that the conditions at the end of the analysis are different from when it started. (Interestingly enough there is also a problem if the feedback is too slow, but that is another matter.) Feedback takes time, and if time is in short supply something must be done to reduce the demands. The solution is feedforward or anticipatory control, which can be defined as acting on an expected deviation or a disturbance but before it has happened. Feedforward effectively reduces the demands on time, because it becomes superfluous to evaluate the feedback in detail. It is, however, only possible to rely on feedforward if the system has a good representation of the process to be controlled, in particular of how it will behave or develop. In practice there are obviously limits to how far a system can rely on feedforward and the most effective strategy is usually a mixture of feedback and feedforward. This means that the feedforward control every now and then is interrupted by an evaluation of the current state, more specifically a comparison of the actual state with the expected state. Any differences lead to a correction – possibly also a correction of the underlying model – after which the system can proceed.

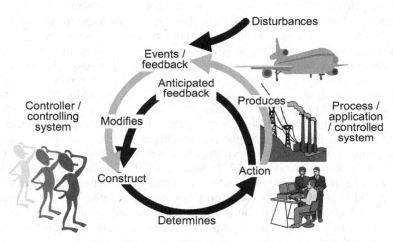

Figure 7.1: Selective use of feedback.

Although feedback and feedforward usually are treated separately, they are in fact intrinsically linked. There is always an element of feedforward in feedback control in the sense that the regulating or compensating action is based on the assumption that it will bring about the desired effect. If that is not the case, the compensating action regresses to trial-and-error, i.e., a random choice of action without knowing whether it will achieve the desired objective. Even if the choice of a compensating action is not based on a deliberation or explicit choice, it may be based on habit (modelled, e.g., by a Markov progression), which essentially means the same, i.e., that the chosen action is more likely to achieve the objective than any other action. Such weighing of alternatives is, of course, feasible only if the environment is relatively stable. If that is not the case, as in the Queen's croquet-ground where the rules changed unpredictably, then one action may be as good – or as bad – as another.

There is also an element of feedback in feedforward. Although the choice of a compensating action is based entirely on assumptions (i.e., the internal model), the carrying out of the action itself requires some kind of feedback. If, for instance, you want to turn on a burner for 45 seconds then there must be something to indicate that the 45 seconds have gone, i.e., some kind of internal feedback. This happens at a lower level than the feedforward, just as in the above case the feedforward happened at a higher level than the feedback.

Deciding on the proper mixture of feedback and feedforward control is not a trivial matter. It is rarely a decision made by the controlling system as such, but rather the outcome of some kind of background process. Neither is it something that can be specified *a priori*. The value of referring to this balance is that it highlights a crucial feature of effective performance, and that it therefore may serve as a guideline both for performance analysis and for system design.

THE SUBSTANCE OF MODELLING COGNITION

Cognitive modelling, which properly speaking should be called modelling of cognition, has willingly accepted as its purpose to account for what goes on in the human mind. This is not just a question of how the details of cognition should be modelled but of how cognition itself should be defined, hence a question about the substance and subject matter of modelling. The common position corresponds to an axiomatic definition of cognition as an attribute of humans, specifically as the information processes that go on in the mind, between input and output. Yet information processing is nothing more than a currently convenient analogy, although the axiomatic definition blankly accepts it as a fact. The axiomatic definition may be contrasted with a pragmatic definition in which cognition is not an attribute of the human mind, but rather an attribute of specific types of system behaviour. According to the pragmatic definition, any system that performs in an orderly manner by showing evidence of goal directed or controlled actions is said to *be* cognitive (but not necessarily to *have* cognition). This means that humans are cognitive, but also that machines or technological artefacts can be cognitive.

On one level it is an irrefutable fact that something goes on between perception and action, or rather as a part of action (since even the 'in-between' view makes a strong assumption about the separateness of process stages). Yet the question is whether we should model that or action itself, and if so whether we should model the actions of the human or the action of the JCS. In general terms, it is necessary to define clearly what should be modelled and how it should be expressed. The first is the issue of the *substance* of modelling while the second is the technical issue of the *language* of modelling.

Cognition without Context

Arguably the earliest early attempt to describe cognition as specific processes or functions was made by Edwin Boring (1946), who outlined a five-step programme for how the functions of the human mind could be accounted for in an objective manner.

1. The functional capacities of the human should be analysed, e.g., by listing the essential functions.
2. The functional capacities should be translated into properties of the organism by providing a description of the input-output characteristics (essentially a black-box approach).
3. The functions should be reformulated as properties of a hypothetical artefact, specifically as a programme for the then emerging digital computer.
4. The properties specified above should be used to design and construct actual artefacts, equivalent to programming the functions in detail and running a simulation.
5. Finally, the workings of the artefact should be explained by known physical principles.

Although written nearly 60 years ago, the approach is basically identical to the principles of information processing psychology, as it became popular in the mid-1970s. The essence of this view was that cognition should be studied and understood as an inner or mental process rather than as action. More particularly, cognition was explained in terms of fundamental processes, hence in all essence treated as an information processing epiphenomenon. This idea of context-free cognition was promoted by Herbert A. Simon and others, who argued convincingly for the notion of a set of elementary information processes in the mind. One consequence of this assumption was that the complexity of human behaviour was seen as being due to the complexity of the environment, rather than the complexity of human cognition (Simon, 1972). This made it legitimate to model human cognition independently of the context, which in effect was reduced to a set of inputs. In the same manner, actions were reduced to a set of outputs, which together with the inputs represented the direct interaction with the context.

Since the main interest of human information processing was to model cognition *per se*, the chosen approach corresponded well to the purpose. CSE, on the other hand, is rather more interested in developing better ways of analysing and predicting performance of cognitive systems. This purpose cannot be achieved by the information processing types of models, and therefore requires an alternative approach.

If we use the axiomatic definition introduced above, the language of cognition refers to descriptions of what is assumed to take place in the mind or in the brain, i.e., the neurophysiological or computational processes to which cognition allegedly can be reduced. If instead we use the pragmatic definition, the language of cognition refers to the theories and concepts that are used to describe the orderliness of performance of cognitive systems and JCSs. In both cases the language of cognition is important because the terms and concepts that are used determine which phenomena come into focus and what an acceptable explanation is. This is illustrated by the classical work in the study of cognition (e.g., Newell & Simon, 1972), which used astute introspection in well-controlled conditions to try to understand what went on in peoples' heads, predicated on the notion of information processes. While many of the features of cognition that have been found in this way undeniably are correct on the phenomenological level, their description often implies assumptions about the nature of cognition that are ambiguous, incorrect, or unverifiable.

Cognition in Context

Since the late 1980s the scientific disciplines that study cognition – predominantly cognitive science, cognitive psychology, and cognitive engineering – have repeatedly emphasised the relation between context and cognition. This has been expressed in a number of books, such as Hutchins (1995) and Klein et al. (1993). The essence of this 'new look', which has been referred to by terms such as 'situated cognition', 'natural cognition' and 'cognition in the wild' is:

- that cognition is distributed across multiple natural and artificial cognitive systems rather than being confined to a single individual;
- that cognition is part of a stream of activity rather than being confined to a short moment preceding a response to an event;
- that sets of active cognitive systems are embedded in a social environment or context that constrains their activities and provides resources;
- that the level of activity is not constant but has transitions and evolutions; and
- that almost all activity is aided by something or someone beyond the unit of the individual cognitive agent, i.e., by an artefact.

Many people have seen this development as a significant step forward, although the initial enthusiasm has not been the same on both sides of the

Atlantic. Yet while it is praiseworthy, indeed, that the study of cognition at long last acknowledges that cognition and context are inseparable, it should not be forgotten that 'situated cognition' is not really something new, as demonstrated by the quotations from Dewey in Chapter 1 and Neisser and Broadbent in Chapter 2. After years of meandering in the *cul-de-sac* of human information processing, it is easy to see the shortcomings of this approach to the study of cognition. It is less easy to see what the problems are in the alternative view, since its power to solve – or rather dissolve – many of the problems that were intractable to information processing is deceptive. Although it was a capital mistake to assume that cognition could be studied without considering the context, it is equally much a mistake to assume that there is a difference between so-called cognition in context, i.e., in natural situations whatever they may be, and so-called context-free cognition. (This is regardless of whether one subscribes to the axiomatic or pragmatic definition.) The methods of investigation may be widely different in the two cases, but this does not warrant the assumption that the object of study – cognition – is different as well.

The hypothetico-deductive approach preferred by academic psychology emphasises the importance of controlled conditions, where independent variables can be manipulated to observe the effects on meticulously defined dependent variables. This classical approach is often contrasted to the so-called naturalistic studies, which put the emphasis on less controlled but (supposedly) more realistic studies in the field or in near-natural working environments. It is assumed that the naturalistic type of study is inherently more valid, and that the controlled experiments run the risk of introducing artefacts and of studying the artefacts rather than the 'real' phenomena.

Many of the claims for naturalistic studies are, however, inflated in a misguided, but understandable, attempt to juxtapose one paradigm (the 'new') against the other (the 'old') and to promote the conclusion that the 'new' is better than the 'old'. Quite apart from the fact that the 'new' paradigm is hardly new at all (e.g., Brunswik, 1956), the juxtaposition disregards the fundamental truth that all human performance is constrained, regardless of whether it takes place under controlled or naturalistic conditions. Given any set of conditions, whatever they may be, some activities are more likely than others – and indeed some may be outright impossible under given circumstances. In the study of fire fighters during actual fires, the goals (e.g., to put out the fire as quickly as possible) and the constraints (resources, working conditions, command and control paths, roles and responsibilities, experience, etc.) will to a large extent determine what the fire fighters are likely to do and how they will respond to challenges and events. In the study of fire fighters using, e.g., a forest-fire simulation game (Waern & Cañas, 2003), there will be other goals and constraints, hence a different performance. The difference between the controlled and the naturalistic situations is not the existence of constraints as such, but rather the degree to which the constraints are pre-defined and subject to control. In either case the performance will be representative of the situation, but the two situations may not be representative of each other.

In the academic study of human performance, the hypothetico-deductive approach requires the conditions of a controlled experiment in order to succeed and the *ceteris paribus* principle reigns supreme, although it is well known that this strong assumption rarely is fulfilled. Indeed, whenever the degree of control is less than assumed, it leads to problems in data analysis and interpretation. The important point is, however, to realise that all human performance is constrained by the conditions under which it takes place, and that this principle holds for 'natural' performance as well as controlled experiments. For the naturalistic situation it is therefore important to find the constraints by prior analysis. If that is done, then we have achieved a degree of understanding of the situation that is similar to our understanding of the controlled experiment, although we may not be able to establish the same degree of absolute control of specific conditions (e.g., when and how an event begins and develops). Conversely, there is nothing that prevents a 'naturalistic' approach to controlled studies, as long as the term is understood to mean only that the constraints are revealed by analysing the situation rather than by specifying it in minute detail. Possibly the only thing that cannot be achieved in a controlled setting are the long-term effects and developments that are found in real life.

Mental Models and the Law of Requisite Variety

One way of resolving the substance issue is to consider the Law of Requisite Variety, described in Chapter 2, which suggests that the study of cognition should focus on problems that are representative of human performance, i.e., which constitute the core of the observed variety. Everyday experience, from academia and praxis alike tells us that the outcome of the regularity of the environment is a set of representative ways of functioning, described by terms such as habits, stereotyped responses, procedures or rules, tactics, strategies, heuristics, etc. We should therefore study these rather than performances that are derived only from theoretical predictions or from impoverished experimental conditions. The substance of cognitive models should be the variety of human performance, as it can be ascertained from experience and empirical studies – but, emphatically, not from theoretical studies. Similarly, in CSE the substance should be the variety of the JCS ascertained in the same manner. The requirement to the model is therefore that its variety is sufficient to match the observed variety of human or JCS performance. The model must, in essence, be able to predict the actual performance of the JCS to a given set of events under given conditions.

The difference between the observed variety and the theoretically possible variety is essential. The theoretically possible variety is an artefact, which mirrors the assumptions about human behaviour and human cognition that are inherent in the theory. The theoretically possible variety may therefore include types of performance that will not occur in practice – because the theory may be inadequate or because of the influence of the working conditions. (It follows that if the working conditions are very restricted, then only a very simple model is needed. This principle has been demonstrated by innumerable experimental studies!) If research is rooted in a very detailed theory or model we may at best achieve no more than reinforcing our belief in the theory. The model may fail to match the observed variety and may very likely also be more complex than strictly needed; i.e., the model is an artefact rather than a veridical model in the sense of being based on observable phenomena.

Using the pragmatic definition of cognition, the requirements to modelling should be derived from the observed variety, since there is clearly no reason to have more variety in the model than needed to account for actual performance. The decision about how much is needed can therefore be based on the simple principle that if the predicted performance matches the actual performance sufficiently well, then the model has sufficient variety. This furthermore does away with the affliction of model validation. The catch is, of course, in the meaning of 'sufficiently' which, in any given context and for any given purpose, must be replaced by a well-defined expression. Yet even if this problem may sometimes be difficult to solve, the answer to the substance issue should definitely be found by asking what cognition *does*, rather than what cognition *is*, in agreement with the pragmatic rather than the axiomatic definition.

COCOM – CONTEXTUAL CONTROL MODEL

In the modelling of cognition, a distinction can be made between procedural prototype and contextual control models. A procedural prototype model assumes that a pre-defined sequence of (elementary) actions or a procedural pattern exists, which represents a more natural way of doing things than others. In a situation, the expected next action can therefore be found by referring to the natural ordering of actions implied by the prototype. A contextual control model implies that actions are determined by the context rather than by an inherent sequential relation between them. In a situation, the next action is therefore determined by the current context and by the competence of the JCS (cf. below). If recurring patterns of actions are found, this can be ascribed to the characteristics of the environment rather than pre-programmed action sequences.

Model Constituents

The Contextual Control Model (COCOM) is a minimal model in the sense that it focuses on the functions deemed necessary to explain orderly performance and is intended to be applicable to a range of JCSs from individuals to organisations. Since such systems may be very different in their substance, the descriptions must necessarily refrain from going into details about the possible underlying structures, except where such an association is obvious. As a minimal model, the COCOM has only three main constituents, called competence, control, and constructs.

- *Competence* represents the set of possible actions or responses that a JCS can apply to a situation to meet recognised needs and demands. The extent of this set depends on the level of detail or the granularity of the analysis, and it is not necessarily denumerable. Furthermore, in terms of modelling, the JCS cannot do something that either is not available as a possible action or that cannot be constructed or aggregated from the available possible actions.
- *Control* characterises the orderliness of performance and the way in which competence is applied. The COCOM deliberately simplifies the description of control to a set of four control modes representing characteristic regions on a continuum going from no control at all to completely deterministic performance. One issue of control has to do with the conditions under which it is either lost or regained, in modelling terms described as changes from one control mode to another. A second issue has to do with the characteristic performance in a given mode – i.e., what determines how actions are chosen and carried out. Both issues are addressed by the COCOM, and define the requirements to the internal functions of the model (Hollnagel, 2000).
- *Constructs* refer to the description of the situation used by the system to evaluate events and select actions. The term is intended to emphasise that this description is a construction or reconstruction of salient aspects of the situation, and that it is usually temporary. Constructs are similar to the schemata of Neisser (1976) in the sense that they are the basis for selecting actions and interpreting information.

An essential part of control is planning what to do in the short-term, within the system's time horizon. This planning is influenced by the context, by knowledge or experience of dependencies between actions, and by expectations about how the situation is going to develop – in particular about which resources are and will be available to the person. The resulting plan describes a sequence of the possible actions, which can either be constructed or predefined. Frequently occurring plans or patterns therefore reflect the relative constancy (regularity) of the environment rather than the constraints of the performance model.

If we refer to the basic principles of the cyclical model (cf., Figure 7.1 above), two essential dependencies can be found. One concerns the revision or development of the construct, i.e., maintaining a correct understanding of the situation. Using the letters E, C, and A to represent events, construct, and actions, respectively, construct maintenance can be described as:

$$C_t \leftarrow E_t | C_{t-1}$$

which means that the construct at time t, the current understanding, is determined by the event given the construct at time $t\text{-}1$, i.e., the previous understanding. This relation provides a link from the past to the present and represents the *reactive* aspects of the model. The time it takes to do this is denoted as T_E or the time needed for feedback evaluation.

The other dependency concerns the selection of the next action – although the term selection does not mean that this in any way is an explicit decision. The selection can be described as follows:

$$A_t \leftarrow C_t | E_{t+1}$$

which means that the action at time t is determined by the current construct given the expected outcome of the action (E_{t+1}). This relation provides a link from the present to the future and represents the *proactive* aspects of the model. The time it takes to do this is denoted as T_S or the time needed to select an action. The temporal relations described by the model will be treated in Chapter 8.

Control Modes

A primary feature of the COCOM is the control modes, which correspond to characteristic differences in the orderliness or regularity of performance. Although the control that a JCS can have over a situation may vary continuously, it is useful to make a distinction between the following four control modes:

- In the *scrambled* control mode, the choice of next action is basically random. For humans there is little, if any, reflection or thinking involved but rather a blind trial-and-error type of performance. This is typically the case when situation assessment is deficient or paralysed and there accordingly is little or no correspondence between the situation and the actions. The scrambled control mode includes the extreme situation of zero control. It can lead to a vicious circle of failed attempts, which is broken when one of the attempts succeeds. In model terms this corresponds to a transition in control mode.

- In the *opportunistic* control mode, the salient features of the current context determine the next action. Planning or anticipation is limited, perhaps because the situation is not clearly understood or because time is limited. An action may be tried if it is associated with the desired outcome, but without considering whether the conditions for carrying it out are met. Opportunistic control is a heuristic that is applied when the constructs are inadequate, either due to lack of competence, an unusual state of the environment, or detrimental working conditions. The resulting choice of actions is often inefficient, leading to many useless attempts being made. Success is determined by the immediate outcome, disregarding possible delayed effects.
- The *tactical* control mode corresponds to situations where performance more or less follows a known procedure or rule. The time horizon goes beyond the dominant needs of the present, but planning is of limited scope or range and the needs taken into account may sometimes be *ad hoc*. If an action cannot be carried out because the pre-conditions are not fulfilled, establishing the pre-conditions may become a new goal (cf. also the description of the goals-means analysis in Chapter 6). The determination of whether an action was successful will take delayed effects into account.
- Finally, in the *strategic* control mode, the JCS has a longer time horizon and can look ahead at higher-level goals. The dominant features of the current situation, including demand characteristics of information and interfaces, therefore have less influence on the choice of action. At the strategic level the functional dependencies between task steps and the interaction between multiple goals will also be taken into account in planning. Outcomes are successful if goal post-conditions are achieved at the proper time and if other goals not jeopardised.

The scrambled control mode is clearly the least efficient, while the strategic is the most efficient – seen from the perspectives of either efficiency or safety. In practice, normal human performance, and therefore also the performance of JCSs in general, is likely to be a mixture of the opportunistic and the tactical control modes. This corresponds to an equilibrium condition or balance between feedback and feedforward and therefore to an efficient use of available resources. Although the strategic control mode theoretically speaking is optimal, it usually requires so much effort that it cannot be sustained for longer periods of time. The main characteristics of the control modes are summarised in Table 7.2.

The COCOM describes JCS performance as a mixture of feedback-controlled and feedforward-controlled activities. This offers a way of capturing the dynamic relationships between situation understanding (as constructs), actions (as realised competence), and feedback or information (as events). On a general level, the model shows how actions depend on the current understanding (construct), which in turn depends on the feedback and information (events) received by the system, which in its turn depend on the actions that were carried out, thereby closing the circle. If control is lost, the actions will likely be incorrect and inefficient, thereby leading to an increasing number of unexpected events. This will necessitate a revision of the construct, which requires both time and effort. If the revision fails, the loss of control will remain and the situation may deteriorate even further. If, on the other hand, the revision succeeds, control may gradually be regained.

Table 7.2: Main Control Mode Characteristics

Control mode	Number of goals	Subjectively available time	Evaluation of outcome	Selection of action
Strategic	Several	Abundant	Elaborate	Based on models/predictions
Tactical	Several (limited)	Adequate	Detailed	Based on plans/experience
Opportunistic	One or two (competing)	Just adequate	Concrete	Based on habits/ association
Scrambled	One	Inadequate	Rudimentary	Random

In the beginning of this chapter, control was defined as the ability to direct and manage the development of events, and especially to compensate for disturbances and disruptions in a timely and effective manner. In terms of control modes this can now be described as the ability to maintain a control mode despite disturbing influences as well as the ability to regain a control mode, should control have been lost. For a given domain it is possible to develop more precise criteria for control mode transitions, and use these as the basis for a model (e.g., Yoshida et al., 1997). The performance characteristics of the control modes can also be used as a basis for identifying the status of a JCS, hence as the basis of performance monitoring systems (e.g., Hollnagel & Niwa, 2001).

Although COCOM is adequate to describe the basic dynamics of control and to illustrate the principle of the control modes, actions can best be understood by invoking different but có-existing layers of control. These refer to different levels of performance rather than to different levels of information processing, hence to a property of the JCS rather than of the operator's internal cognition.

ECOM – EXTENDED CONTROL MODEL

In order to describe multiple levels of performance it is necessary to extend the basic COCOM model, thereby turning it into an Extended Control Model (ECOM). ECOM provides a way of describing how the performance of a JCS takes place on several layers of control simultaneously, corresponding to several concurrent control loops. Some of these are of a closed-loop type or reactive, some are of an open-loop type or proactive, and some are mixed. Additionally, it acknowledges that the overall level of control can vary, and this variability is an essential factor with regard to the efficiency and reliability of performance.

The idea that human behaviour comprises multiple simultaneous processes is far from new. An early allusion to that was provided by Craik (1943), as well as by Miller, Galanter & Pribram (1960). As noted in Chapter 4, MacKay (1968) also made the same point early on, though in a more general manner; a later cybernetic model is the recursive hierarchy in the Viable Systems Model (Beer, 1981; 1985). Broadbent (1977) provided an excellent overview of the issues and pointed out that multilayered models must comprise independent processes at each layer. Carver & Scheier (1982) proposed a five-level system, although with feedback loops only. Other well-know examples are Michon (1985) and Rasmussen (1986). Most of the information processing models, however, fail on Broadbent's criterion since they do not involve parallel processes of a different nature but only the same type of process repeated at different layers.

As far as the number of layers goes there is obviously no absolute reference to be found. Since the purpose is to provide a way to describe, analyse and model actual performance, the number of layers should be sufficient to serve this purpose, but not so large that the descriptions become unmanageable. Using the stages in the development of process control as an analogy, a useful distinction can be made between four layers or loops called tracking, regulating, monitoring, and targeting (goal-setting), respectively. Adopting these four layers for the description given here does not in any way preclude a revision at a later point in time. The assumption of multiple layers of activity is crucial for the modelling approach, but the specific number of layers or loops is not. In the following sections, each of the four layers of activity will be described using the example of driving a car.

Tracking

The tracking loop describes the activities required to keep the JCS inside predetermined performance boundaries, typically in terms of safety or efficiency. In the case of driving a car this could mean maintaining an intended speed, a safe distance from cars around, etc. Activities at the tracking layer are very much a question of closed-loop control. For the skilled user such activities are performed automatically, without paying much attention to them, and therefore with little effort. (Note that this applies only to experienced users. For a novice these tasks are not a question of tracking but rather of regulating, cf. below.) While activities at the tracking layer usually are performed in an automatic and unattended manner, they may become attended, hence more like regulating, if conditions change. (Consider, for instance, driving on a snowy or icy road. Here the normally simple activity of keeping the lateral position of the car – or just keeping a steady course – suddenly becomes very demanding.)

In ECOM, goals and criteria for activities at the tracking layer are provided by the regulating layer. Most of the tracking activities are furthermore amenable to technology take-over and automation. In the case of driving a car, speed can be maintained by cruise control and in some cases longitudinal separation distance may be maintained by adaptive cruise control. In the case of vessels and aeroplanes, both direction and speed are under automatic control. The automation of activities at the tracking layer often give rise to automation surprises. This is because the almost complete take-over of closed-loop control makes it very difficult for the operator to follow what is going on, hence to maintain the situation comprehension that is needed at the other layers of activity.

Regulating

Activities at the tracking layer require that actions and/or targets (as well as criteria) already exist. In ECOM, these come from the regulating layer, which provides input in terms of new goals and criteria. Regulating is itself basically a closed-loop activity, although anticipatory control may also take place (e.g., McRuer et al., 1978). Activities at the regulating layer do not always take place smoothly and automatically, but may require attention and effort. These activities in turn refer to specific plans and objectives that come from the monitoring layer.

In driving, activities at the regulating layer are concerned with the position of the car relative to other traffic elements, such as avoiding obstacles and relative positioning, e.g., in overtaking, hence multiple tracking subloops. In relating to other traffic, the regulating loop may suspend the tracking loop. It may, for instance, be more important to keep the position in the traffic flow than to maintain a given speed, hence braking (or acceleration) may overrule keeping a steady speed. This can also be expressed as a temporary suspension of one goal (keeping a constant speed) to the advantage of another (maintaining a safety distance to other cars). The incompatibility between goals can be resolved by changing plans, e.g., first to overtake the car in front and then to go back to maintaining a steady speed.

Monitoring

Whereas activities at the regulating layer may lead to either direct actions or goals for the tracking loop, activities at the monitoring layer are mainly concerned with setting objectives and activating plans for actions. This can involve monitoring the condition of the vehicle, although this has in most cases been taken over by instrumentation and automation. Modern cars usually have self-diagnosing systems that only inform the driver in case of serious limit transgressions or malfunctions, and measurements have therefore effectively been replaced by alarms. (For driving the exception is the monitoring of the fuel reserve, which is one of the few measurements that still are provided. In some cars the distance that can be driven using the remaining fuel may also be shown. This can be useful in planning and goal setting.) Other monitoring activities have to do with the location of the vehicle. Whereas position refers to the vehicle relative to other traffic elements, location refers to the vehicle relative to reference points in the environment.

In other domains and processes a similar distinction between regulating (position) and monitoring (location) can be made, although the space may not be a physical (or an Euclidean) one. In a vehicle, other activities at the monitoring layer may have to do with infotainment and information sources. Although this is not monitoring of driving *per se*, it may affect the ability to drive, particularly if it is non-trivial. Monitoring does not directly influence positioning of the vehicle in the sense of closed-loop control and regulation but rather is concerned with the state of the joint driver-car system relative to the driving environment.

Targeting

The last type of action occurs at the targeting or goal setting layer. An obvious kind of goal-setting is with regard to the destination. That goal may give rise to many subgoals and activities, some of which can be automated or supported by information systems. Other goals have to do with criteria for acceptable performance. For instance, if a user expects that the destination will be reached too late, criteria for the other loops, notably regulating and tracking, may be revised. When time is short, the style of driving may change by increasing the speed, reducing the separation distance, and in general taking greater risks. Another example is if the car includes a sensitive consignment, such as a frail person or a delicate piece of equipment. In this case both the driving style and the route may be changed, as avoiding shaking and bumping becomes more important than, e.g., speed or fuel consumption.

Goal-setting is distinctly an open-loop activity, in the sense that it is implemented by a nontrivial set of actions and often covers an extended period of time. Assessing the change relative to the goal is not based on simple feedback, but rather on a loose assessment of the situation – for instance, the estimated distance to the goal. When the assessment is done regularly it may be considered as being a part of monitoring and control (such as in the Test-Operate-Test-Exit loop introduced by Miller, Galanter & Pribram, 1960). If the assessment is done irregularly, the trigger is usually some unknown factor, perhaps time, perhaps a pre-defined cue or landmark (physical or symbolic), perhaps the user's background 'simulation' or estimation of the general progress (like suddenly feeling uneasy about where one is).

ECOM Structure and Parameters

ECOM follows the same principles of minimal modelling as the COCOM and is indeed built on the latter in the sense that each layer corresponds to the fundamental construct-action-event cycle. One way of representing that is as in Figure 7.2, which shows the relations among the four layers in a simplified manner. To avoid graphical clutter, Figure 7.2 includes only the goal dependencies among the layers; other dependencies also exist, for instance, in the propagation of feedback or events. For the same reason each layer is represented only by one construct-action-event cycle, even though there normally will be several, as noted in the text above. Figure 7.2 also indicates the relative weight of feedback and feedforward control for each layer.

The main characteristics of each of the four layers have been described in the preceding text, and are summarised in Table 7.3. This shows for each layer the type of control involved, the typical demands to attention in the case of a human controller, how often events can be expected to occur, and finally the typical duration of events. Here it is important to note that events on the tracking layer usually are of such short duration that in the case of humans they are pre-attentive. In other words, tracking-type behaviour is equivalent to skills, in the sense that it is something that is done more or less automatically and without attention. The regulating layer comprises actions of a short duration that do require attention, but not for long. The monitoring layer describes actions that go on intermittently as long as the task lasts, although the distribution can be decidedly irregular, depending on demands. Finally, actions on the targeting layer take place every now and then, almost always including the preparation of a task.

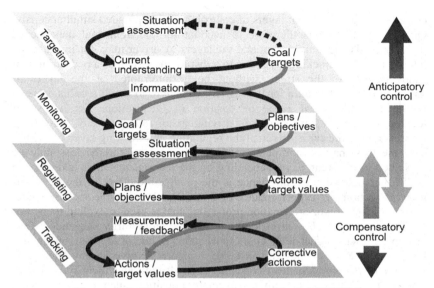

Figure 7.2: The Extended Control Model (ECOM).

Table 7.3: Functional Characteristics of ECOM Layers

	Control layer			
	Tracking	Regulating	Monitoring	Targeting
Type of control involved	Feedback	Feedforward + feedback	Feedback (condition monitoring)	Feedforward (goal setting)
Demands to attention	None (pre-attentive)	High (uncommon actions); low (common actions)	Low, intermittent	High, concentrated
Frequency	Continuous	Medium to high (context dependent)	Intermittent, but regular	Low (preparations, re-targeting)
Typical duration	< 1 sec ('instanta-neous')	1 sec – 1 minute ('short term')	10 minutes – task duration ('long term')	Short (minutes)

Interaction between Loops

The above description shows how ECOM can be used to describe some of the interactions between the different loops. The assumption throughout is that all loops are active simultaneously, or rather that goals and objectives corresponding to different layers of control are being pursued simultaneously. One use of ECOM is therefore to account for the nontrivial dependence between goals and activities among the layers. A driver may, for instance, be interrupted at the tracking layer due to a disturbance, such as a pedestrian that suddenly crosses the street, without being interrupted at the layers of targeting or monitoring: a local evasion manoeuvre to avoid a collision need not lead to a loss of sense of position or localisation. Yet a loss of localisation, such as when one is driving in an unknown city, may have effects on how the car is driven at the regulating and tracking layers. In either case the loss may be abrupt or gradual. The reasons for the loss may be explained in different ways, depending on the theoretical stance taken. From a CSE perspective, the importance is the predictable effect of the between-loop interaction on the performance of the JCS, rather than the specific model of micro-cognition or information processing that is used to explain it.

The goals of each control loop can also be temporarily suspended. One example is that a higher-level goal is suspended in lieu of focusing on a lower level one. The driver may, for instance, temporarily give up any attempt to get to a location and instead concentrate on driving and identifying where s/he is (meaning absolute rather than relative position, such as when driving in an unknown city). The driver may also suspend the regulating and tracking loops or goals, e.g., by pulling up to the curb, or stopping the car.

The bottom line is that controller performance and the effect of support systems can be understood only in the context of the JCS. The importance of the environment can be seen in the relative importance of different goals. In urban traffic, for instance, the driver who is familiar with the streets will navigate from waypoint to waypoint in a closed-loop manner, paying attention mostly to the flow of traffic, although often in a highly automated manner without full awareness. The attention may instead be on listening to the radio, planning the activities of the day, or simply flicker around without any recognisable purpose. Driving can be smooth and efficient because little effort is required to monitor the position relative to the destination. For the inexperienced driver the situation is quite different. Monitoring the location becomes very important and more effort is assigned to it. The actual driving, though still a closed-loop activity, may therefore take place at a slower speed and with lower efficiency. For both types of drivers, open road driving is more consistent, since less effort is needed to control the position amongst other traffic elements, and more effort can be used to monitor the progress towards the destination.

Modelling the Loss of Control

Finally, a few comments on the relation between ECOM and the COCOM are in place. At present, COCOM can be seen as an elaboration of the basic cyclical model with emphasis on the different control modes, i.e., how control can be lost and regained, and how the control modes affect the quality of performance. ECOM does not change this significantly. The degree of control can still be considered relative to the layers of ECOM.

On the tracking layer the activities are performed in an automated and unattended manner. Even if a person runs away from a fire in a state of panic, the running itself is unaffected – or may even be improved! It therefore makes little sense to talk about a loss of control, although the tracking loop clearly may be disrupted by external events. On the other three layers (regulating, monitoring, targeting) control can clearly be lost, leading to a degradation of performance. The loss of control is very much a question of losing goals, or of selecting inappropriate goals and criteria. Relative to COCOM, ECOM should not introduce any significant changes in the description of layers of control, and the performance cycle (selection-action-evaluation) remains as a unit of description. ECOM does not imply that there are simultaneous cycles as such, but rather that the formulation of goals is described on several layers, i.e., a simultaneity or concurrence of goals and intentions. In terms of human performance it is possible to do only two things at the same time if one of them is on the tracking layer, i.e., if it is automated and unattended. The apparent ability to do several things at the same time is due to the ability to switch or share between goals in an efficient manner. ECOM thus describes the relation between multiple layers of goals, rather than between multiple layers of action as such.

Chapter 8

Time and Control

Since CSE is about the control of dynamic processes and environments, time is of fundamental importance. Time plays a role in coping with complexity as well as in handling data and information. The chapter describes how time is an integral part of the CSE models, as well as the various techniques and strategies that can be used to overcome a possible shortage of time.

ORTHODOXY IN MODELLING

When it comes to the practice of modelling, two characteristic but widely different perspectives can be recognised. The first is focused on the *how* of modelling and therefore concerned mainly with the structure and contents of models. It takes the need of modelling for granted and proceeds to work out the most effective or elegant ways in which this need can be fulfilled. This type of modelling is structural in the sense that the main interest is in the architecture of the model, i.e., what the components are and how they are configured. Important distinctions relate to types of models (e.g. mathematical, computational, analogue, symbolic, etc.), and the components that go into the models. In the case of modelling human operators it has been common to include components as required by a specific theory, such as the conventional model of decision making, the standard model of human information processing (involving different types of memories and attention resources), etc.

The second perspective is focused on *what* is being modelled and is concerned mainly with the functioning or performance of the model. This approach begins by clarifying exactly what the need of modelling is, hence determining the meaningfulness of the question before beginning to answer it. Such modelling is usually functional, in the sense that the main interest is what the model does, rather than how it is done. The purpose and the functioning of the model are therefore more important than the structure or contents of the model. To take a simple example, the fact that retention of information is limited is important in most cases and must therefore be part

157

of the model. In a structural model the common solution is to include a component such as a short-term memory. Yet short-term memory basically expresses a functional characteristic, which can be modelled in other ways (Landauer, 1975). In the functional approach to modelling there are no requirements for models to include specific components; i.e., models are not built bottom-up. Instead, the most efficient way of representing or reproducing a specific function is chosen, corresponding to a phenomenon driven principle.

The issue of orthodoxy in modelling is perhaps easier to see if we consider it at the level of how a structural model is implemented. Assuming that a specific model architecture has been agreed upon, it can be implemented by a computer in many different ways using a variety of representations and formalisms. As long as the model implementation behaves or functions as required, it matters little in which language it is written and under which operating system it is executed. Indeed, it is easily realised that since the model is just computer code, and since the choice of code is arbitrary relative to what is being modelled, there is little reason to be overly concerned about that or to require that the model be structured in a specific way or that it contains specific components. Yet the same line of argument can be applied to the model itself. The requirement is to have a model that behaves or functions in a certain manner, i.e., which reproduces or simulates specific behavioural characteristics. Whether this is done by using the components of the conventional information-processing model, or by applying a different principle, should be of little concern. Although the information processing view is firmly entrenched in present-day psychology, human factors engineering, and cognitive science, it is still a culturally determined perspective rather than a law of nature. The introspective corroboration of information processing in the mind is not independent proof, but rather a consequence of applying the terminology in the first place.

To argue this point even further, consider how philosophers and psychologists described what we now call human cognition before the invention of the computer, hence the availability of the computer metaphor. People have clearly thought, solved problems and made decisions in the same way for thousands of years, which is why it is still relevant to read Aristotle. We need to go back only around 120 years to William James' *The Principles of Psychology* (James, 1981, org. 1890), to find descriptions of human cognition that still are valid, yet without a trace of information processing. Indeed, James's description of attention is still unrivalled by later developments (cf., Norman, 1976).

Model Minimization

Warnings about the dangers in an unreflecting use of modelling by Neisser (1976) and Broadbent (1980) have already been mentioned (Chapter 2). The latter provided an extensive discussion of what he termed "the minimization of models" and made the plea that "models should as far as possible come at the end of research and not at the beginning" (Broadbent, 1980, p. 113). The alternative to start from a model would be to start from practical problems and concentrate on the dominant phenomena that were revealed in this way, corresponding to Neisser's (1976) notion of natural purposeful activity. Models would still play a role in choosing between alternative methods of investigation and in supporting the interpretation of data; but models should in all cases be minimized to avoid that choices became driven by the model rather than by practical problems.

The rationale underlying these pleas is that the regularity of the environment leads to a set of representative ways of functioning, and that we should investigate these rather than performances that are derived only from theoretical predictions or found in impoverished experimental conditions. If this advice is followed the resulting models will be minimized in the sense that they focus or concentrate on essential phenomena. Another way of addressing this issue is by considering whether a model truly is the simplest possible description of the target phenomenon or whether it rather expresses the assumptions of an underlying computational or psychological theory.

It is, however, not unproblematic to find the representative ways of functioning, partly because what we see depends on what we expect to see. Perception, whether part of research or just managing your daily life, is not the neutral registration of what is 'out there', but is an active filtering of what can be found. This problem has been recognised from the earliest days of psychology and described by Peirce (1958; org. 1868) as the problem that "every cognition is determined logically by previous cognitions". It therefore behoves the researcher carefully to consider which preconceptions s/he may have, so as to minimise their effects. Although there clearly is no need to have more variety in a cognitive model than required to account for the observed variety, being successful in having the right amount of variety, hence being able to model or control the observed variety, does not mean that the model captures the true nature of human performance or provides a true explanation of cognition. The model may be pragmatically correct and pass the test of verifiability (in the sense that counterexamples cannot be produced) but still not be true in the sense of offering an intellectually, theoretically or even philosophically acceptable explanation or set of concepts. This touches upon the nature of proof as a whole, cf. the discussion in mathematics of whether proof is a social or a logical phenomenon (Colburn, 1991). The Law of Requisite Variety has an undeniable flavour of 'rightness' because it works. It also has the advantage that models based on this principle may remain applicable despite changes in cognitive psychology and cognitive science; in contrast, a model derived from a theory that currently is in favour will likely be invalidated by a paradigm shift.

Two Neglected Issues

Following Broadbent's advice we therefore need to determine the requisite variety that should be modelled. This is fortunately rather easy to do, at least as a first approximation. One observation is that human performance is not always homogeneous, but that it can be more or less orderly or well regulated (Hollnagel, 1998b). Another is that thinking and acting – and, indeed, everything else – both *take time* and *take place in* time. Both of these aspects are so obvious that they seem to be taken for granted, at least in the sense that the common models of human information processing or human cognition have not addressed them. Consider, for instance, the widely used description of human performance as being either skill-based, rule-based, or knowledge-based (Rasmussen, 1986). The aspect of control is absent despite the fact that the model purports to describe the differences between three levels of processing. While the levels differ in terms of the types of performance they represent, the individual functions such as feature formation, identification, and recognition seem to be carried out with uniform regularity regardless of variations in task demands, time pressure, experience, quality of the interface, etc. Similarly, the aspect of time is absent from the model, both in terms of the time taken by the cognitive functions to work and in terms of the time demands of the work environment, which for this type of model mainly relates to the dynamics of the process that is being controlled.

Control and time are nevertheless indispensable aspects of human action, not only as individual features but also in how actions are coupled and organised. As everyday experience shows, control may easily break down when time is too short. A good model of human performance should therefore be able to account for the dependency between performance and time. This is furthermore not the only aspect to consider; other candidates are affects or emotions, motivation, learning, etc. It can easily be argued that each of these is more important for human performance – and therefore also for how it is modelled – than, e.g., short-term memory, attention, workload, etc. They are unfortunately also more difficult to reconcile with the information processing paradigm (e.g., Simon, 1967), which may partly explain why they have been neglected. In this book we will restrict ourselves to focus on the aspects of control and time, since some progress already has been made with respect to these. Yet this does not imply that the other aspects are any less important.

MODELLING CONTROL

Chapter 7 introduced the Contextual Control Model (COCOM) and with that the four control modes: strategic, tactical, opportunistic, and scrambled. The control modes were introduced because it is important for a model of human performance to be able to replicate the operator's dynamic selection of actions, the various ways in which this can take place, and the influence of internal and external information such as the operator's current interpretation of the state, previous actions, time and resource constraints, and external factors (communication).

The notion of control modes gives rise to several important research problems. One problem is the characteristics of performance for a given control mode and, of course, how such performance itself may affect future control modes. Another problem concerns the conditions that may lead to a change in control mode – keeping in mind that the control modes refer to characteristic regions on a continuum, hence that they are continuous rather than discrete. If the intention is to develop an actual contextual control model, for instance as a computer simulation, these problems must be answered in detail. For the present discussion it is sufficient to outline some of the main relations, as shown in Figure 8.1, namely how well the operator (or in general, the controlling system) is able to predict events, and how much time is available.

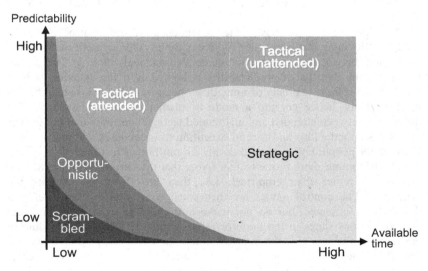

Figure 8.1: Main determinants of control modes.

The first thing to notice from Figure 8.1 is that there are five rather than four control modes, because a distinction is made between a tactical attended and a tactical unattended control mode. This is in recognition of the fact that people often seem to relax control of what they are doing when the situation is not demanding. Consider, for instance, the following description of the relation between attention and performance:

> We see that highly learned, automatic activities may require little or no conscious attention to be performed. These highly automated activities can be disrupted, and when that happens they require conscious control, often to the detriment of their normal performance. (Norman, 1976, p. 70)

Highly learned, automatic activities are typically carried out in recurring, hence familiar situations. This is almost the same as saying that predictability is high; people therefore resort to routine and very often drift into doing other things at the same time. The classical example of that is driving to work by the normal route. While doing that, there is time to listen to the radio, talk on the phone or plan the day ahead. The advantage is that there is spare capacity; the disadvantage is that the reduced attention may lead to mistakes or reduced reliability, cf. Hollnagel (1993a) or Reason & Mycielska (1982). The familiarity of the situation and the lack of time pressure will in most cases reduce the thoroughness of what is being done. Because of this, superficial cues may trigger responses that in the situation are incorrect, such as making a turn.

As Figure 8.1 also shows, the dependence of the control mode on predictability and available time is not simple. This is reflected by the distinction between attended and unattended tactical control. The model represents the fact that if predictability is high, i.e., if the situation is very familiar and if there is plenty of time, then operators will pay less attention to what happens, hence go into a mode of unattended tactical control. The difference between attended and unattended tactical control is that the former represents a meticulous and careful execution of procedures and plans, while in the latter people know what to do but do not bother to follow it in detail, thereby becoming more prone to fail. (Note that the curves in Figure 8.1 are theoretical, rather than empirical, i.e., they are there to illustrate the principles. The control modes are theoretical constructs that are useful to describe performance, but do not make any claim to exist in reality, and certainly not as states in the brain or mind. Figure 8.1 is also misleading in the sense that it may give the impression that the two axes, predictability and available time, are independent of each other. This is, however, not the case. For instance, if predictability is low, then it may take more time to think through a situation, which means that less time remains available for other things. Conversely, if there is little time available, then predictions may be harder to make.)

The main features of Figure 8.1 are summarised in Table 8.1.

Predictability

As pointed out in Chapter 7, human performance relies on a mixture of feedforward and feedback control. Feedforward control is essential to prevent performance from becoming purely reactive with little or no opportunity to consider the situation as a whole, hence to predict or plan ahead. In human behaviour, feedforward is equivalent to anticipatory control, which requires that the operator is able correctly to anticipate what will happen and prepare himself/herself to act accordingly. Indeed, even in so fundamental a phenomenon as perception, anticipation plays a crucial role. It stands to reason, that the farther – and more correctly – predictions can be made, the better will the operator be able to maintain control of the system. The ability correctly to anticipate future events and developments depends on a number of things, such as knowledge and experience, quality of information available, designed support such as procedures, the regularity of the process and environment, etc.

Table 8.1: Control Mode Characteristics

Control mode	Available time (subjective)	Familiarity of situation	Level of attention
Strategic	Abundant	Routine or novel	Medium – high
Tactical (attended)	Limited, but adequate	Routine, but not quite – or task is very important	Medium – high
Tactical (unattended)	More than adequate	Very familiar or routine, almost boring	Low
Opportunistic	Short or inadequate	Vaguely familiar but not fully recognised	High
Scrambled	Highly inadequate	Situation not recognised	Full (hyper-attention)

Available Time

The other main influence on the control mode is the available time. Since more will be said about this in the following, it is sufficient for now to note that when there is too little time available, then it will be very difficult to make predictions (since that requires time), and perhaps difficult even to respond to what happens. A shortage of time will impede feedforward control and may even degrade feedback control. Conversely, if there is ample time then it will be possible for the operator to consider in more detail actual events and potential developments, and plan to act accordingly.

In addition to predictability and available time, there are other conditions that may determine the control mode. One of these is how well the situation is understood. If it is difficult to understand what is going on, for instance because the information presentation is less than optimal, then control may be lost even if there are no or only a few unexpected events. For the operator it is necessary to be able to understand what happens, i.e., the present status of the process, as well as what has happened and what may happen. What has happened corresponds to the short-term or long-term past of the process, while what may happen refers to the future. Knowledge of the past is essential in order to be able to recognise conditions, diagnose events and identify disturbances. Assumptions about future developments are necessary in order to plan what to do, to decide on effective means of intervention, and to make decisions about future actions.

THE MODELLING OF TIME

Time refers to the fact that we are dealing with systems and processes that develop and change. The primary application area of CSE is situations, where the rate of change of the process or target system is appreciable relative to the duration of the activities under consideration, in particular situations that change so quickly that it is a problem for the JCS to keep pace. (Second to that, but much less studied, are situations where the rate of change is so slow that it, almost paradoxically, is difficult for the JCS to sustain attention.) The rate of change in turn depends on how the system boundaries have been defined, and there are therefore no permanent criteria for when a situation is dynamic or not. This means two things: first, that there is limited time available to evaluate events, to plan what to do, and to do it. Second, that the information that is used needs to be updated and verified regularly because the world is changing. This is one reason why it is unrealistic to describe decision making as a step-by-step process unless the decision steps are minuscule relative to the speed of the process.

If unexpected events occur occasionally, there may be time and resources to cope with them without disrupting the ongoing activities, i.e., without adversely affecting the ability to maintain control. But if the unexpected events are numerous and if they cannot be ignored, they will interfere with the ongoing activities, hence lead to a loss of control. This is a result of the simple fact that it takes time to deal with unexpected events. Since time is a limited resource, the net effect will be that there is less time, hence less capacity remaining to do the main tasks. Such situations may occur for all kinds of JCSs, ranging from single individuals interacting with simple machines to groups engaged in complex collaborative undertakings. Indeed, it soon becomes evident that, regardless of domains, a number of common conditions characterise how well they perform, and when and how they lose control.

Time has generally been treated as a Cinderella in both human-computer interaction and human-machine interaction (Decortis & De Keyser, 1988). This is also the case for cognitive engineering and cognitive science, despite the obvious importance of time in actual work, i.e., in activities that go on outside the controlled confines of the laboratory. The proximal reason for that is probably the legacy from behaviourism, carried on by human information processing psychology, which focused on how an organism responded to a stimulus or event, rather than on how an organism or system behaved over time. The distal reason is the fundamental characteristic of the experimental approach to scientific investigation, whether in the behavioural or natural sciences, which is to expose a system to an influence and take note of the consequences or reactions (Hammond, 1993). It is thereby assumed that the behaviour being studied can be decomposed into discrete chunks without affecting its functional characteristics.

While it has been known since the days of Donders (1969, org. 1868-69) that mental processes take time, the speed of actions is more important than the speed of mental processes. In other words, the interesting phenomenon is the time it takes to do something, such as recognising a situation or decide about what to do, rather than the time of the component mental processes, particularly since these cannot be assumed to be additive except for specially created experimental tasks (Collins & Quillian, 1969). One simple reason is that it cannot be assumed that the duration of an action – to the extent that one can talk about this in a meaningful way at all – can be derived by considering the duration of the elementary or component processes. Even if the internal workings of the mind were sequential, in a step-by-step fashion, the combination or aggregation need not be linear. Human action is furthermore not the execution of a single sequence of steps, but rather a set of concurrent activities that address goals or objectives with different time frames and changing priorities. For example, in order to make decisions, a process plant operator needs to be able to reason about temporal information and changes, to predict the effects of his actions and of the changes he observes, continuously to make reference to what has happened, is happening and might possibly happen, and to co-ordinate on the temporal axis the actions of several users (Volta, 1986).

To illustrate the vast differences in temporal demands that any serious study of JCS has to consider, Table 8.2 shows the profile of a number of different domains and tasks (adapted from Alty et al., 1985). The domains range from the daily and almost trivial (cycling and driving a car) to highly specialised and demanding activities (flying an aeroplane or controlling a power station). The entries are arranged according to the maximum number of process variables. As Table 8.2 shows, this ranges from the very few to the exceedingly many; other domains, such as grid control, may run even higher. Two other important aspects are how often it is necessary for the operator to do something and how much time is available to do it. The frequency of operator operations almost falls into two separate groups, one where actions are required very frequently if not continuously, and one where actions are few and far between. The time allowed category is also very interesting, since one domain stands out from the others, namely nuclear power generating stations. The 30-minute respite to act found here is a deliberate solution to a serious problem, as discussed in Chapter 4.

Table 8.2: Temporal Characteristics of Different Domains

Process type/domain	Number of process variables	Frequency of operator actions	Time allowed for operator actions
Cycling	2 (speed, direction)	1/second (manoeuvring) 1/minute (coasting)	Direct
Car driving	< 10	1/second (heavy traffic) 1/minute (light traffic)	Direct
Steel rolling mills	< 100	1/second	Direct
Aviation	100 - 300	1/minute (landing) 2-3/hour (cruising) 1/second (manual flight)	Direct
Electronic trading	~500 – 5,000	1-4/minute	<< 1 minute
Process industries	2,000 – 10,000	5-6/hour (sometimes clustered)	< 1 minute
Nuclear power generating stations	10,000 – 20,000	1/hour (usually clustered)	1-30 minutes

Representation of Time in COCOM

Effective control requires that the operator – and more generally, the JCS controlling the process – is able to make sense of the available information and in particular possible unexpected events, as well as able to find, choose or generate appropriate actions or responses. The model shown in Figure 8.2 comprises several sets of time that affect the ability to remain in control. One set comprises the time needed for various parts of an action, i.e., the durations of the components of an action. These are the time needed to evaluate events and update/develop an understanding of the situation (T_E), the time needed to choose or select an appropriate action (T_S), and the time window allowed for execution after the response has been selected (T_W). (A more complete version of the model also includes T_P, the time needed to perform an action.)

The other set comprises the points in time when things happen, denoted as τ_O and τ_{LFT}. The former is the point in time when an action becomes necessary, either due to an external event or due to the person (e.g., an intention or recognised need to do something). The latter is the latest finishing time (LFT) for the action that corresponds to what happened at τ_O (cf. Allen, 1983). This means that the total time available to respond to an event cannot be larger than $T_A = \tau_{LFT} - \tau_O$. In practice, the time available to carry out the response is T_W rather than T_A, where $T_W = T_A - (T_E \cup T_S)$. (We use $T_E \cup T_S$ rather than $T_E + T_S$ because T_E and T_S may be overlap.) According to this line of reasoning, if $(T_E \cup T_S \cup T_P) > T_A$ then the operator has too little time to understand what is going on and to respond effectively, and control therefore will sooner or later be lost. Conversely, when $(T_E \cup T_S \cup T_P) < T_A$, then the operator has enough time to understand what is going on and to choose and effectuate a response, and is therefore likely to remain in control and possibly even able to plan ahead. (The simplicity of the model in Figure 8.2 suggests that T_E and T_A occur in a sequence and at discrete times. That is, however, an artefact of the graphical representation. In reality, event evaluation and action choice will usually be intermingled, although there is a logical necessity that the former precede the latter. Figure 8.2 represents the case of a single action of undefined scope. However, it is quite possible to apply the same type of reasoning to more complex interactions and to combined or aggregated actions although it becomes messier. The principles are nevertheless easiest to illustrate with an idealised case.)

Research on human-computer interaction has generally referred to tasks that were *self-paced* and where limited time therefore played only a minor role. In contrast to that, tasks in most industrial domains are *force-paced* or process-paced. The available time, T_A, is determined by the speed of the process and if $(T_E \cup T_S \cup T_P)$ exceeds T_A, it puts severe constraints on the operators' possibility to evaluate events and select actions. Some processes, such as steel rolling mills, electronic trading, or flying an airplane, require rapid or even near instantaneous responses (cf. Table 8.2). Other processes such as power generation, land-based transportation or surgery pose less severe demands but still require that actions be taken within a limited time. It would clearly be better if there were ample time, i.e., if $(T_E \cup T_S \cup T_P)$ were less than T_A, since the operator then would have time not only to respond in the situation but also to refine the current understanding, to plan before acting, hence to improve control of the situation. This can be achieved if the time limitations can be relaxed, for instance by slowing down the process, although that is possible in only very few cases. A more common approach is to reduce either T_E or T_S by improving the system and interface design, although this usually is done in a piecemeal fashion.

In cases where the process develops at a regular or stable speed and where there are few unexpected events, the τ_{LFT} of $event_i$ is predictable, since it corresponds to τ_O of the following $event_{i+1}$. There are, however, many cases where the progress of events is irregular or where unexpected events are common. Unexpected events are not just due to possible failures of the process, but equally and more often due to the unpredictability of the environment (cf. the discussion in Chapter 7). If τ_{LFT} is uncertain, then the JCS in principle has two possible strategies. One is to stick to the task and ignore unexpected events, unless they are very serious. (Note that this does require some kind of quick evaluation or screening, similar to the problem in selective attention.) The other is to try to complete the action at hand as quickly as possible, i.e., to minimise T_E, T_S, and T_P. This is consistent with the ETTO principle, according to which people trade-off thoroughness for efficiency in order to gain time. In this case the trade-off is, however, due to uncertainty rather than to external production pressures.

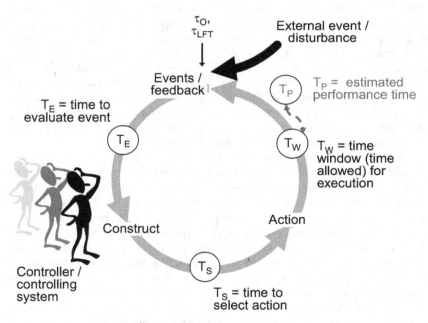

Figure 8.2: Time and control in COCOM.

Time and Control Modes

The predominant position of event evaluation and action selection in COCOM provides an easy way of accounting for the coupling between time and control and thereby to provide a little more detail to the description of Figure 8.1. As described elsewhere (Bye, Hollnagel & Brendeford, 1999), event evaluation and action selection can be carried out to various depths depending on the control mode. This corresponds to the basic fact that humans, unlike machines, may be more or less thorough in what they do depending, among other things, on the available time. This provides both a single way of expressing the time-control dependency conceptually and a simple way of implementing it computationally.

To study this in more detail it is useful for a moment to depart from the cyclical model and focus exclusively on the different time sets. The relations can be laid out as shown in Figure 8.3, which depicts a situation where there is only a single line of action. It is therefore allowable to treat the events as if they were ordered sequentially. Figure 8.3 illustrates the problems that can occur when the three prototypical parts of an action have to be accommodated within an available time window, T_A. In cases where the available time is insufficient for event evaluation, action selection, and execution together, the JCS will be incapable of responding before it is too late. If this condition is prolonged, it will sooner or later lead to a loss of control, hence to a degradation of the control mode. The relations can be described as in Table 8.3.

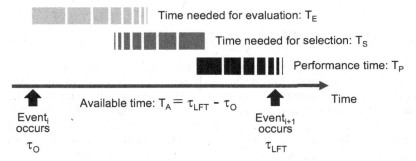

Figure 8.3: Temporal relations at work.

The conditions listed in Table 8.3 are indicative rather than exhaustive. This goes especially for the condition corresponding to the opportunistic control mode. One obvious improvement would be to add a conditional condition, for instance that $(T_E + T_S + T_P) < T_A$. Another would be to increase the resolution by considering the preparation and the execution of the action separately, e.g., as $[(T_E + T_S) < (T_A - T_P)]$ OR $(T_P < T_W)$. Yet another improvement, and an important one, would be to recognise that the various times referred to here are approximate rather than precise, hence introducing, for instance, $Min(T_E)$ and $Max(T_E)$ as the upper and lower boundaries for T_E, respectively. This could be even further refined by applying the categorisation of time proposed by Allen (1983).

For the present purpose the descriptions and proposals above are, however, sufficient to show that it is possible to speak about the temporal characteristics of actions in an unambiguous manner, and that it furthermore is possible to propose qualitative – and perhaps also quantitative – relations between time and control. In reality the situation will always be more complex than the examples used here, and it is an abstraction to speak about time without in the same breath mentioning the conditions on which available time depends. The main conditions are the type of process (process domain), the state of the process (e.g., normal or disturbed), the frequency of unexpected events in the process or environment, the appropriateness (or quality) of previous actions and the anticipation of feedback, and the time used for previous actions.

Table 8.3: Control Mode Dependency on Time

Condition	Resulting control mode
$(T_E \cup T_S \cup T_P) > T_A$	Scrambled
$(T_E > T_A)$ XOR $(T_S > T_A)$ XOR $(T_P > T_A)$	Opportunistic
$(T_E \cup T_S \cup T_P) < T_A$	Tactical
$(T_E \cup T_S \cup T_P) \ll T_A$	Strategic

HOW TO ENHANCE CONTROL

A shortage of time is always a problem, but it is fortunately a problem to which several solutions can be found. It is, indeed, possible to see many common design features as ways of reducing temporal demands. The strategies or tricks used to compensate for a shortage of time can conveniently be discussed under two headings, one in terms of the technical solutions available and the other in terms of the heuristics that people use to cope with the temporal complexity.

In the following, COCOM will be used as a frame of reference to describe a number of ways in which the problem of time, or rather the problem of a shortage of time, is being addressed. This should not in any way be seen as a validation of the model, but rather as an indication of the usefulness of it. Since too little time is one of the reasons why control can be lost, the various solutions can consequently be seen as ways in which control may be facilitated.

Technological and Organisational Solutions

Control rooms broadly speaking, including aircraft cockpits and others, are awash with technologies that in one way or the other are supposed to make the operator's life easier. Some of the commonly applied solutions are shown in Figure 8.4.

Time to Evaluate Events (T_E)

The time needed to evaluate events, (T_E), can be reduced primarily by improving the design of information presentation. This has for many years been one of the dominating concerns in the field of human-machine interaction, and a number of high-level design principles have been put forward, including adaptive displays (Furukawa & Inagaki, 1999), multimedia interfaces (Bergan & Alty, 1992), and ecological interface design (Vicente et al., 1995). The lofty goal has been to provide the operator with 'the right information, in the right format, at the right time', although this in most cases is easier to say than to do (cf. the discussion in Chapter 4).

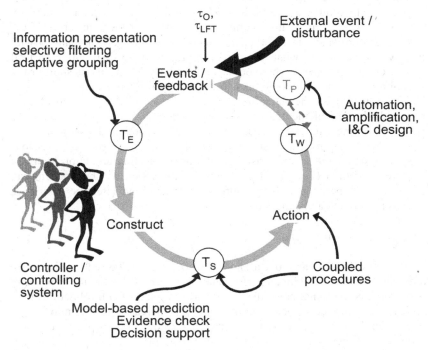

Figure 8.4: Technological solutions to alleviate time shortage.

Another common solution is suppression of information, either by filtering information according to logical principles or by using sophisticated graphical information presentation, such as adaptive grouping or task-oriented displays. Other proposals involve computer-supported interpretation and diagnosis, expert systems, and artificial intelligence in various forms. These, however, have the drawback that they begin to introduce automation, the problems of which were discussed in Chapter 6. Less technology dominated solutions to the problem of reducing T_E primarily involve education, training, and better use of human resources such as in Crew Resource Management techniques.

Time to Select Actions (T_S)

The primary means to reduce the time needed to select an action, T_S, is by means of procedures and extensive training. Procedures in effect encapsulate prepared decisions, which relieve operators of having to go through the reasoning behind the procedure. The time to select actions can be reduced by improving the structure and formatting of procedures, or even by computerising procedures (e.g., Rouse, 1981), although this may be a double-edged sword. Another approach is to improve the human-machine interface, not so much in terms of information presentation but more in terms of the design of controls surfaces and panels, whether hard or soft. In analogy with the 'what you see is what you get' mantra of HCI, CSE might instead propose 'what you see is what you do'. Particularly when time is short or the situation is unfamiliar, people tend to be heavily influenced by the looks of the controls and by the functions they afford access to. Both efforts must obviously be complemented by training, which in itself can improve the ability to estimate the time needed to carry out an action, T_P, as mentioned above. More sophisticated solutions involve various forms of decision-support systems and computerised predictions. Selecting an action logically involves a prediction of what the outcome of the action will be, and this can in many ways be supported by appropriate system and interface design, as well as by training and experience.

Time to Perform an Action (T_P)

The time needed to carry out a chosen action, T_P, can be reduced by using automation to amplify human performance. The use of automation is in many ways a mixed blessing, since the speed and precision of control actions may be offset by reduced observability and an increase in the number of unexpected events – leading to the dreaded automation surprises and new problems of when to intervene (Dekker & Woods, 1999). Invoking additional resources, such as extra staff, may also help to reduce the performance time although there is an obvious trade-off involved. On the organisational scale rapid deployment forces can reduce T_P if it is worth the additional cost.

Available Time

Finally, it may under some conditions be possible directly to increase the time available (T_A). In some cases the speed of the process, hence the occurrence of τ_{LFT}, can be reduced. For surface transportation, for instance, the speed can be reduced – in the extreme case to a complete standstill. For many other processes, the speed of the process usually is beyond human control or can be changed only within very narrow limits, but several other solutions are possible such as the 30-minute rule discussed above. For organisational processes another solution is to renegotiate deadlines, for instance in software engineering or large construction projects. This requires, however, both that the whole development can be slowed down without irreparable damage and that there is some degree of power to influence the environment.

Human Solutions

In addition to solutions that require technological or organisational resources, individual humans also resort to a number of ways to overcome a temporary shortage of time (Figure 8.5). We have already discussed some of these under the heading of *Coping with Complexity* (Chapter 4).

Time to Evaluate Events (T_E)

The strategies described in Chapter 4 can be used to specify how COCOM should respond to an IIO condition. Since the overall objective of the various strategies is to reduce the time needed to deal with the incoming information (i.e., minimize $T_E \cup T_S$), coping strategies must be considered both for event evaluation and action selection. Two of the strategies, *decentralisation* and *escape*, can be eliminated directly since neither is realistic. The remaining strategies are conceivable either for event evaluation, for action selection, or for both. Omission, for instance, would in terms of event evaluation correspond to a pseudo-random sampling of the available information. Another example is cutting categories, which for event evaluation would entail using only part of the available information. Both strategies are possible and, indeed, likely. Queuing and filtering would also work, although the former merely postpones the problem. Other common human strategies to reduce the T_E are to use approximate rather than precise principles of evaluation, such as similarity matching and frequency gambling (Reason,

1990), or the representativeness and availability heuristics (Tversky & Kahneman, 1974).

Action selection is essentially a form of decision making, and the speed of that can be improved by several simplifying strategies such as satisficing (Simon, 1955) or 'muddling through' (Lindblom, 1959). Another way of reducing T_S, and in many cases also T_E, is to rely on pattern matching or recognition. Amalberti & Deblon (1992), for instance, found that fighter pilots put considerable effort into planning and preparing a mission, precisely because this reduced the need of decision making during flights. This is practically indistinguishable from the strategy of recognition-primed decision making described by Klein et al., 1993. Another way of reducing the demands to time is to make a trade-off between efficiency and thoroughness, as described by the ETTO principle (Hollnagel, 2004).

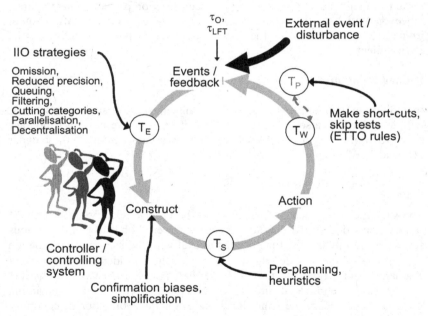

Figure 8.5: Human solutions to alleviate time shortage.

Time to Perform an Action (T_P)

For the human operator, T_P can in principle be reduced by working faster, although this might not be very effective in the long run. One reason is that it may lead to more action failures; another that the speed of the process may make it risky to go too fast. The Therac-25 accident (Leveson, 1993) is a vivid example of that. T_P can also be reduced by changing the task, i.e., task tailoring, but now due to time demands rather than to system quirks. A reduction of T_P is often achieved by skipping steps of test and verification, as described by such ETTO rules as, 'it usually works', 'someone will test this later', or 'someone has already tested this'. The risks inherent in this strategy are obvious and can indeed be found as part of many accident explanations.

Available Time

Since time is determined by the nature of the process, there is not much that people can do to increase the available time, apart from the strategies that were mentioned under the technological and organisational solutions.

CONCLUSIONS

Any consideration of human performance, and specifically human performance in complex technological systems, soon makes it clear that the broad performance characteristics may be more important than the minutiae of information processing mechanisms or cognition in the mind. This chapter has considered two of them: that actions take place in time and are affected by time, and that actions are carried out with varying degrees of control.

It is difficult to reconcile these characteristics with the common variety of human information processing models, since these rarely consider time and control. The conventional procedural prototype models are pure feedback and response models in which actions are determined more by the characteristics of the internal mechanisms than by the conditions of work. Human performance is, however, proactive as well as reactive and therefore requires a model that describes actions as determined by the context rather than by a pre-defined order relation between constituent functions.

Contextual control models can be used to describe important characteristics of how a system is able to perform in an orderly manner, and to do so regardless of whether that system is an individual human being, a technological artefact, a socio-technical system, or an organisation (Worm, 2001). COCOM thus represents an approach to modelling, rather than a specific model. This can be seen from the fact that the principle of contextual control modelling has been implemented in several different ways, although with mainly the same purpose.

One example is a mathematical model for the cognitive performance of a tactical air traffic controller (Blom et al., 2001), which was used in an evaluation of accident risks for en-route controllers. The model combined the features of a general human error model, Chris Wickens' Multiple Resource

Model, and COCOM. Corker & Verma (2000) used COCOM to account for the role of context in the study of air traffic management and free flight. This work implemented the concept of control modes and how they depended on the context of operations as the basis for a third generation version of the Man-Machine Interactive Design and Analysis System (MIDAS). In a different domain Bye et al. (1999) used COCOM as one of the main components in an analytical study of function allocation between operators and automation systems in Nuclear Power Plant (NPP) control rooms. The two other components were a model of the automation systems and a model of the nuclear power plant. A more formal proof-of-principle was attempted by Yoshida et al., (1997) who developed a system to make real-time recognition of operator control models, using the principles of the contextual control model. Finally, Stanton et al. (2001), used COCOM to categorise team behaviour in terms of the four control modes, and demonstrated that it was possible to determine the control mode with a high degree of confidence.

These applications of the principles of contextual control models at least demonstrate the viability of the concept, and show that it is entirely feasible to develop models of user performance that are minimal in the sense that Broadbent (1980) argued. The purpose has in all cases been to develop a model of the operator that can be used for something, i.e., a practical version of the machine's image. This represents a general shift from structural towards functional modelling that is evident in many areas, including cognitive ergonomics and accident analysis. It should be expected that the parts and composition of the contextual control model, such as they are, may change as experience accrues. What must remain, however, is the basic objective to develop models of operators that capture the requisite variety of human performance in a specific environment.

Chapter 9

CSE and Its Applications

This chapter provides a brief overview of the actual and potential applications of CSE. These include work and task design, control room design, decision support, etc. The problems are, of course, well known, since they stem from the practical problems of working life and of human-machine systems. CSE provides a consistent approach to address these problems, as well as a set of methods with a common conceptual basis.

WHAT SHOULD CSE BE ABOUT?

The preceding chapters of this book have identified the main practical and research issues of CSE and introduced the basic concepts, theories, and models. This last chapter will provide a brief overview of both actual and potential applications of CSE. These include task and work design, system and interface design, automation, accident and risk analysis, task design, etc. None of the problems are, of course, unique to CSE since they stem from the practical problems of working life and of human-machine systems. But CSE provides a consistent approach to address these problems, as well as a set of methods that have a common conceptual basis. It is, however, impossible to go into any real detail of the applications of CSE in a single chapter. This chapter therefore provides a high-level introduction, while details must await a separate treatment at another time.

The study of JCSs in context is a process of discovering how the behaviour and strategies of practitioners develop to meet changing purposes and constraints in a field of activity. The key is to uncover how people learn to exploit capabilities and resolve dilemmas and conflicts as they pursue their goals. Since general patterns in work do not exist separate from specific intersections of people, technologies and work, studies must be carried out in field settings. To do so requires ways of discovering the essential patterns underlying the diversity of cognition and of going behind the more or less visible activities. The main field settings that have been studied by CSE include the alarm problem in tightly coupled systems, how people respond to anomalies, the role of diagnosis in disturbance management, the problems with automation (in particular clumsy automation, escalation and automation surprises), common ground in coordinated activity, oversimplifications in coping with complexity, and behaviour under conditions of data overload. Many of these issues have already been introduced in earlier chapters as the practical problems that drove the development of the theories and models of CSE. In many cases they also define the challenges for CSE in the years to come.

The lessons learned from trying to solve these problems have had a significant influence on CSE as it has gradually matured and have, as often as not, emphasised the need to be extremely careful in the definition of what the significant aspects of the practical problems are. Following the advice of Broadbent, Neisser, and others to start with the problems rather than with the solutions, this has progressively resulted in something that, with a bit of hesitation, may be called the basic methodological principles of CSE.

The first principle is carefully to identify situations where problems exist, or predict situations where problems may arise, with the emphasis on the reality of the problems. Identifying the nature of problems is not always as easy as it seems, with the sad story of the search for 'human error' being a case in point. After much ado it turned out that the problem was not one of 'human error' *sui generis*, but rather one of understanding how well-intended and sensible actions every now and then could lead to unwanted consequences. In consequence of that it might therefore be wise to replace the vacuous term 'human error' with something like a concept of Actions With Adverse Consequences (AWACS). It is a small comfort that other branches of behavioural science have experienced similar problems in searching for 'mythical' entities, such as Lashley's search for the engram (Lashley, 1950), and the theories of memory (Jenkins, 1974).

The second principle is to describe the conditions associated with the problems, either as potential causes or as factors that affect how an event develops. The emphasis here is on the external conditions rather than on the assumed mechanisms or internal processes. Before trying to explain anything in terms of its definitive causes, it is necessary to understand fully the context where the problems were observed. The Western belief in the determining powers of the human mind, whether in normal work or psychopathology, constitutes a bias that often leads us to search in the wrong direction.

The third and final principle is to propose or construct the means by which such situations can be mitigated or prevented. This emphasises that it is more important for CSE to produce practical solutions than to develop elegant theories and models. There is certainly a high degree of intellectual satisfaction in being able to explain why something has happened, but the real value lies in being able to turn the explanation into practical methods. There is, unfortunately, no shortage of examples of problems that have been 'solved' by applying readily available methods without first understanding what the nature of the problem was. (The area of automation can provide countless examples of that, e.g., Woods et al., 1994.) Just as problems often have been technology induced, so have solutions equally often been technology driven. (The area of user interface design is a good illustration of that, for instance the early fervour for multi-media and virtual reality.) Yet we should not try to solve the problems for which we have methods, but rather try to develop methods for the problems that we find.

Extensions to Human Factors/Ergonomics

The changing nature of work must be matched by a development of models and methods that can be used to describe how artefacts become an integral part of work, as well as to design them. Both human factors engineering and ergonomics were developed to deal with the problems of the day and did so in a highly satisfactory manner. The changes to the nature of work during the last 50 years have, however, been of a character and magnitude that require considerable changes to models and methods. Yet as we have argued above, the approaches to the description of human-machine systems have for many years seemed to be irrevocably stuck in the information-processing paradigm, specifically in the deceptive decomposition of systems into humans and machines.

The increasing complexity of work environments has nevertheless forced a reassessment in several disciplines, for instance accident analysis (Hollnagel, 2004). The study of human behaviour has also, after some hesitation, changed its flavour from a study of cognition in the mind to a study of cognition in the world (Hutchins, 1995; Klein et al., 1993). Yet the conventional disciplines of human factors engineering and ergonomics have not undergone a similar change. This has led to a predicament because they are unable to address the issues that other disciplines and potentially rivals – such as cognitive ergonomics, human-computer interaction, computer supported co-operative work (CSCW) – successfully deal with. Although the predicament perhaps is yet to be fully acknowledged, it is quite obvious that the approaches to the design of work systems or JCSs must be broadened considerably. One suggestion for a possible development is given in Figure 9.1, which shows how the current basis for dealing with work systems must be extended along two dimensions.

The first dimension refers to the section of the system's life cycle that is being considered. The emphasis has traditionally been on the operation of the system or the artefact, i.e., the actual use of the artefact for the designated purpose, also known as the sharp end. It is becoming clear that systems must be considered over a much longer time span – beginning with system design and ending with maintenance, or perhaps even decommissioning. Problems of operation cannot be understood without considering also how the JCS or artefact was designed with all the considerations that went into that. Many of the problems met at the sharp end have their origin in choices made during the design phase, and many others have their origin in activities that take place during maintenance. This is specifically the case for complex industrial systems, not just because they are large and complex, but also because there is considerable pressure to keep the maintenance phase as brief as possible – for the simple reason that the system is inoperative during maintenance. (One solution to this problem is to perform maintenance on-line, i.e., without taking the system out of operation, although this carries with it appreciable risks.)

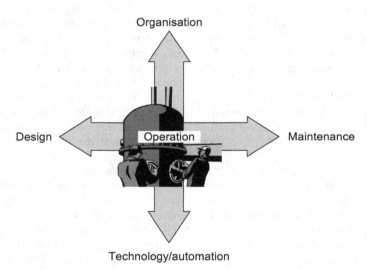

Figure 9.1: Extensions to human factors/ergonomics.

The other dimension refers to the scope of system functionality being considered. Again, the traditional view of operations at the sharp end has focused on what one might call the proximal influences on performance. There is, however, a need to consider the distal influences as well. These can be found buried either deep inside the technology, e.g., in the form of automation, or in the organisational environment in the form of rules, principles, and tacit practices. It is the ambition of CSE to provide the conceptual and methodological basis for such an extended view, i.e., both in scope and time. The remaining pages of this chapter will illustrate how this can be done in practice using three significant issues: the design of work, control room design, and decision support.

DESIGN OF WORK (COGNITIVE TASK DESIGN)

The primary target of design is usually the direct interaction with or use of the artefact – as in ergonomics and human-computer interaction/human-machine interaction. This follows from the structural view of a system in terms of its parts and the connections among them. Yet such an approach has two shortcomings. The first is that it considers the system as decomposed rather than as a whole, as a JCS. The second is that it concentrates on how artefacts are used but pays scant attention to the planning and organising that in most cases is a prerequisite for proper use. Because of this, traditional work design is oblivious to the fact that the use of an artefact involves the latter as much – or even more – than the former.

After an artefact has been in use for a period of time, long enough for people to have reached a plateau of performance, the consequences of use can be seen both in the direct and concrete (physical) interaction with the artefact as well as in how the use or interaction with the artefact is planned and organised. This is obviously the case for technological artefacts and information devices, since these directly affect user tasks. A little thought makes it clear that the same is true for any kind of design or intentional change to a system, since the use of the system, i.e., the way in which functions are accomplished and tasks carried out, will be affected. Introducing a new 'tool' therefore affects not only how work is done, but also how it is conceived of and organised. This will in most cases have consequences for other parts of work, and may lead to unforeseen changes with either manifest or latent effects. The design of work must accordingly go beyond interface and interaction design, and consider not only short term effects at the sharp end, but also medium and long term effects (cf. Chapter 6). All design of artefacts, whether technological or social, or of JCSs, is therefore also inevitably the design of cognitive tasks (Hollnagel, 2003).

The focus of Cognitive Task Design (CTD) is on describing the performance of JCSs in the complex socio-technical networks that provide the foundation of our societies, and how this performance must keep changing to enable the systems to stay in control. CTD therefore struggles with the dilemma known as the envisioned word problem (Woods, 1998), i.e., how the results of a cognitive task analysis that characterises cognitive and cooperative activities in a field of practice can be applied to the design process, since the introduction of new technology will transform the nature of practice! Or put more directly, the paradox of CTD is that the artefacts we design change the very assumptions on which they were designed.

Making Work Easy

There are many aspects to work design that require consideration. Task analysis was originally developed to improve the efficiency of physical work (Taylor, 1911). The goals of classical ergonomics have throughout been issues of a very tangible nature, such as how to reduce occupational injury and illness, how to improve productivity, how to improve work quality, or how to reduce absenteeism. In more recent times, usability has been considered essential to bridge the often self-created gap between human and machine (Nielsen, 1993).

Since CSE is about maintaining control of work, two overall perspectives more or less suggest themselves. One is how to make work easy, in the sense of making it easy to maintain control. The other is how to make work safe, in the sense of minimising the possibilities for unwanted outcomes, as well as minimising the potential effects or consequences of such unwanted outcomes. We shall briefly look at each of these issues in turn.

Making work easy can be seen as an issue of designing facilitators for work, cf. the discussion of amplification in Chapter 2. Most aspects of interface and interaction design can be seen as providing facilitators, i.e., as ways of making it easier for users to understand and use the system correctly. On a high level, work facilitation can be considered from the following perspectives.

- Good interface design. A golden rule here is to ensure that basic ergonomics and human factors principles are observed both with regard to information presentation and the exercise of control. Although there are many guidelines and standards available (e.g., Smith & Mosier, 1986; Helander et al., 1997), it is not uncommon to find blatant violations of both guidelines and established cultural stereotypes.
- Simplicity of functioning. When trying to use an artefact, the strongest rule is 'what you see, is what you do'. In other words, people tend to follow the perceived affordances of the artefact; conversely, hidden or obscure functionality is unlikely to be used. It is therefore important to avoid conflicts between perceived and actual functionality.
- Clearly indicate system operational state. This is a principle that often is violated. It is crucial for the use of automation and systems that have multiple modes that the actual mode is indicated. Failure to do so has contributed to some of the more spectacular accidents but can even befuddle quite simple tasks such as switching a device or an artefact on or off.
- Clear and precise instructions. Instructions can be explicit, such as in the case of procedures, or implicit, such as in the case of artefact design (interface). Instructions can be immensely useful if they are precise, but

too often a mismatch between the instructions and the artefact creates unnecessary complications.

As an example, consider the design of HCI for a start-up procedure of a power plant boiler. The whole procedure consists of 32 different steps of which only six are executed as single steps, i.e., without any parallel or simultaneous activities going on. Figure 9.2 shows a subset of the generic procedure (Lind & Larsen, 1994) describing the step 'raise boiler pressure to starting value for turbine'. This requires some way of exercising control ('raise boiler pressure ...'), as well as a way of reading the pressure and comparing it to the reference value ('... to starting value for turbine'). Depending on the speed of the process one might even wish to include an indication of the trend or rate of increase of the pressure. It is thus obviously a step that involves several HCI design considerations, and also one which can go wrong in several ways. If, for instance, attention is wavering, if the controls are incorrectly timed, or if a step is missed, the top-level goal may not be reached. To design an appropriate HCI for this task therefore requires careful consideration of what the operator is likely to do under different circumstances, in particular of what may happen if the procedure is not executed exactly as expected.

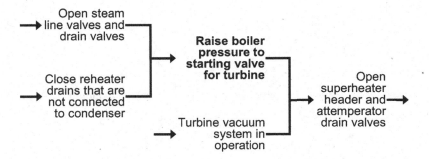

Figure 9.2: Start-up procedure for plant boiler (part).

Making Work Safe

Making work safe comprises two different aspects, cf. Figure 9.3. One is to ensure that work is safe by preventing failures from taking place, possibly even eliminating opportunities for failure or for doing things incorrectly. This can be done either directly by making it more difficult to do something wrong, or indirectly by making it easier to do something right (i.e., facilitation). The other is to try to protect against the consequences of failures, if and when they happen. Here, as everywhere else, prevention is obviously better than cure.

Finding out what could go wrong is, in one sense, relatively easy because there are only a limited number of ways in which performance can fail. These are variously referred to as failure modes, error modes, or phenotypes (Hollnagel, 1993c). Without going into too many details, it stands to reason that actions can go wrong with regard to the following:

- timing (too early, too late),
- duration (too long, too short),
- speed (too fast, too slow)
- direction (too far, too short, wrong direction),
- magnitude (too large, too small),
- force (too much, too little),
- sequence (omission, jumps),
- type (incorrect type of action),
- object (wrong object), and
- location (object is in wrong place).

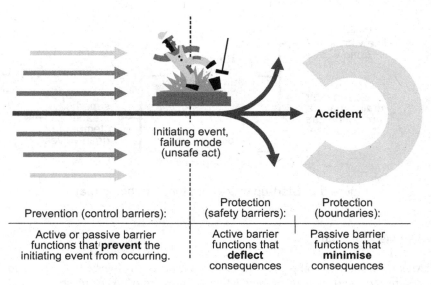

Figure 9.3: Prevention and protection.

Knowing what could possibly go wrong is useful, since it can direct the search for how it specifically can happen. Part of a system design should therefore be to ask questions about whether an action can fail with regard to timing, to duration, to speed, etc., in the same way as when doing a risk analysis. In the engineering domains a number of methods have been developed to help identify the possible risks. These were initially limited to technological failures, represented by such methods as fault trees, event trees, and Failure Mode and Effects Analysis (e.g., Taylor, 1993), but have gradually been enhanced to be able to consider also failures attributable to human actions. This developed into a specific discipline known as Human Reliability Assessment (or Human Reliability Analysis), e.g., Hollnagel, 1993a; Kirwan, 1994. Another development was to consider software failures in more detail, as illustrated by Leveson (1995).

Knowing *how* something can go wrong is, unfortunately, not the same as knowing *why* it can go wrong. Each failure mode may have a number of possible causes. Consider, for instance, possible reasons for hitting the wrong key on a computer keyboard or a mobile phone keypad. Even for a small system like this most people can easily think of a number of different reasons why 'wrong object' may happen (the same goes for the other failure modes). Speculating about why actions could fail, or more specifically why people sometimes made 'errors', created a large subfield of 'human error' studies. Much of this work was based on the assumption that humans could be described as information processing systems, which sometimes could fail, i.e., an endogenous view of 'human error'. In contrast to that it was also possible to see 'human error' as due to a coupling among internal and external factors, hence a more exogenous view.

Over time, the latter view has gained strength and recognition. Neither the endogenous, nor the exogenous view is, however, consistent with CSE. Here the view is rather that 'human errors' represent a *post hoc* rationalisation (Woods et al., 1994), which is based on the inverse causality principle: 'if there is an effect, then there must have been a cause'. CSE instead suggests that we cannot understand what happens when things go wrong without understanding what happens when things go right. We therefore need theories and models of normal actions rather than of 'human error'. Chapters 7 and 8 have proposed that both normal performance and failures should be explained in terms of how people adjust their actions to achieve an acceptable compromise between resources and demands. Efforts to make work safe should therefore start from an understanding of the normal variability of human and JCS performance, rather than assumptions about particular, but highly speculative 'error mechanisms'.

The other part of making work safe is to protect against unexpected and unwanted consequences. This is generally done by including various barriers or defences as parts of the JCS. Despite the obvious importance of this issue, it has been rather neglected in system design – and especially in interface and interaction design. Some considerations of protection have been addressed in the study of occupational health and safety, although aimed mostly at the individual worker. There is, however, still a significant need of research and development that looks into making work safe from an overall systemic perspective.

Problems in the Design of Work

There are two principal issues in the design of work. One is that of having the right model. The other, and related, is of making assumptions explicit. We have already briefly mentioned the issue of having the right model, e.g., as in the distinction between model and image. We may here apply the Law of Requisite Variety once again, since it simply says that the model that is used as basis for the design (of the JCS) must be rich enough, i.e., have enough variety, to correspond to the variety of the real world.

While it seems straightforward to talk about a user model, there are in fact several such models involved in system design (cf. Figure 9.4). There is (1) the designers' ideas about the artefact and the functionality it shall provide (designers' model of system); (2) the designers' ideas or assumptions (cf. below) of who the users are, what they want to do, and what they are capable of doing (designers' model of users); (3) the condensed assumptions that are built into the artefact, i.e., how the artefact actually looks and functions, as different from what the designers intended it to look and function (artefact's or system's image); and finally (4) the end users' ideas about the artefact, what they intend to use it for and how they expect it to function (users' model of system).

There are obviously fairly rich opportunities for mismatches or discrepancies among the four types of models described here, particularly between (3), as a reflection of (1), and (4). Another potential mismatch is between (2) and the actual user characteristics. For CSE it is more important to consider the possible discrepancies and their consequences, rather than the models as such. Indeed, the models are only abstractions or tokens for ideas. To quote Conant & Ashby (1970) yet again, a user does not need to *have* a model of the system but must by necessity *be* a model, i.e., behave or perform in a way that is appropriate to maintain control.

One way of bringing the importance of the hidden design assumptions up front is to consider what the design would look like if the user was an extra-terrestrial alien. (This, by the way, is not as far-fetched as it may seem. The two Voyager spacecrafts that were launched in 1977 carried with them a gold-plated copper phonograph record in an aluminium jacket on which was engraved the instructions for use (Sagan, 1980). Others have for years been trying to send messages to a possible extra-terrestrial intelligence by means of radio.) In order to design the system's interface we must make assumptions about the alien's sensitivity to, e.g., sound, light, or other electromagnetic radiation – which in turn requires the assumption that these modalities are meaningful in themselves. We must make assumptions about the ability to distinguish figure and ground (in any modality), to recognise and understand signs and symbols, to remember configurations and patterns, etc.

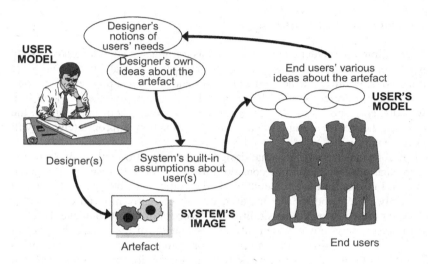

Figure 9.4: Varieties of models.

Many of these assumptions are also made in the design of terrestrial systems, but few of these are explicitly recognised. In the case of a terrestrial alien, it is useful to try to imagine in how many ways a person can be different from oneself. A terrestrial alien might, for instance, have poor eyesight, be partly deaf, have poor motor coordination, not understand the language or terminology, have limited short-term retention, be unable to focus or maintain attention, etc. In that sense we are all aliens at some time – when we are very tired, when the environment is hostile, when time is short, when we are ill, etc. (e.g. Peddie et al., 1990). In such situations actions are more likely to go wrong, but the probability of this happening can be reduced if designers take the necessary precautions.

A terrestrial alien may also be a novice, i.e., someone who is not familiar with the system or who uses it only infrequently. Everyone will obviously be a novice at one time or another, since there is no such thing as a population consisting only of experienced users. More particularly, users rarely have the skill and knowledge of designers or that designers hope for. The set of implicit assumptions is not only large but is also necessary. No system design could set out all the assumptions that are used, just as no system could be designed for a complete alien, i.e., a being or a creature that has nothing at all in common with us. Much of the art of good system design, and in particular good JCS design, consists of finding or defining the right set of assumptions, i.e., the assumptions that will not be invalidated or contradicted by the actual system working conditions or by what actually happens when the system is put into operation.

The Forced Automaton Analogy

Every complex artefact can be described as a finite state automaton or a state machine, in terms of a set of inputs, outputs, internal states, and state transitions. In order for the artefact to work and produce a pre-defined output, it must get a correct input. This means that the operator must respond in a way that corresponds to one of the pre-defined categories of input. If not, the artefact, hence the joint system, will not function appropriately; it will become unreliable.

A consequence of this is that designers are forced to think of operators as automata, because they must consider how operators can respond to a given output. In order for the artefact to work, the output must be such that it can be correctly understood by the operator, i.e., be correctly mapped onto one of the pre-defined response possibilities. If the operator fails to achieve that, i.e., if the response is not included in the expected set of responses, the artefact will eventually malfunction. In practice, operators may interpret information in a nearly infinite number of ways, but it is not possible to account for this in the design of the system. Instead, the designer is forced to consider a finite (and usually rather small) number of interpretations and the reactions that may follow from that. The design, in fact, tries to make the operator respond as a finite automaton, and it is not far from that to actually thinking about the operator as an automaton.

The most obvious example of this approach to design is found in the so-called user-friendly graphical interfaces that have become the standard on all personal computers. It may seem a good idea to provide people with a limited number of choices, as menu items or icons, but it effectively forces them to behave like automata. The situation may even be 'improved' by restricting the selection e.g. by disabling parts of a menu or by establishing logical interlocks (the extreme example is a message box with a single 'OK' button; here the operator can do only one thing). Such limitations will nominally improve the reliability of the HCI because it is impossible for people to do something that has not been anticipated by the designer. It does not mean, however, that the machine always responds as people expect, because they may have formed their own interpretation of what goes on beneath the graphical surface.

By limiting the possibilities the designer really limits the responses, and the user friendly interface becomes a strait-jacket. The use of the system is ostensibly made easier, but only for anticipated situations. It becomes difficult or impossible to explore how the system works, to develop new ways of working (unless they comply with options that the designer has considered), or in general to deviate from what the system allows. The user-friendly systems thereby actually serve the purposes of the designer rather than the purpose of the operator, and the design is made on the premises of the designer rather than on the premises of the operator. In the short term reliability is increased because variety is reduced in the way it manifests itself. However, in the long term reliability can be improved only by understanding the nature of the variety and use that as the basis for design decisions.

CONTROL ROOMS

We have already in Chapter 4 discussed display design and the principle of designing for complexity and noted that it is possible to reduce the complexity of a system, hence also the complexity of the interaction, only if the system boundaries can be changed. However, in most cases where display and interaction design are being considered, it happens within existing system boundaries. The hope of simplifying or reducing the complexity is therefore an illusion as discussed above.

As pointed out in Chapter 4, designing for complexity involves designing for predictability and providing sufficient time. Chapter 8 provided a detailed discussion of the relation between time and control, as well as the ways in which temporal demands can be reduced to increase the match to the available time.

The design of control rooms, and more generally the design of information presentation, is often restricted to considering the information about the present, i.e., the state of the process here and now. It is, however, clear from the discussion in Chapters 7 and 8, that information must be provided about the past, the present, and the future alike – although to varying degrees depending on the nature of the situation (and the state of the process).

A control room can be defined in general as a *location* designed for an *entity* to be in *control* of a *process*. This definition contains a number of key concepts, each of which must be defined on its own. The *location* is simply the place or places where the controlling entity is, and where the input and output devices necessary and sufficient to maintain control can be found. In some cases a control room is a physical room in a physical building. In other cases it is a place (a location) that moves with the process being controlled, e.g. the driver's environment in a vehicle or the cockpit of an aeroplane. Notice that as more than one location can be involved, the definition comprises centralised as well as distributed control. The *entity* is in our case a cognitive system or a joint cognitive system. The meaning of *control* is to minimise or eliminate unwanted process variability, as we have discussed throughout this book. Control therefore has a purpose and hopefully leads to a desired outcome (a goal state). Finally, the *process* is a continuous activity or set of functions. The process has its own dynamics hence changes if left alone.

From these definitions a control room can also be described as a room with a view, where the view is to the past, the present, and the future. The view to the past is necessary in order to understand the current situation, to build up expectations, and to anticipate what may lie ahead. The view to the past provides a general frame of reference so that the operator does not come in 'cold' in a situation, which may make it difficult to understand what is going on.

The view to the present requires little justification. It is obviously necessary to know what is going on, what the current status of the process is, whether that is achieved via a global assessment – a holistic impression – or via specific measurements and indicators (symptoms, crucial variables).

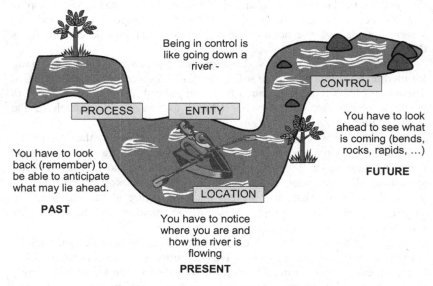

Figure 9.5: "A room with a view".

The view to the future in all essence corresponds to the expectations that follow – logically or otherwise – from the understanding of the system, i.e., of the construct. Having a view of the future enables us to be effective in looking for and sampling information as well as in interpreting it. The downside or risk is of course that this may also lead us astray, for instance ignoring certain information or misinterpreting it. To take a slightly frivolous example, process control can be likened to going down a river in a boat (Figure 9.5). The 'control room' location is the boat; the entity is the person in the boat; the process is the flowing river with its bends and rocks; and control is the ability to negotiate potential risks in going down the river, such as avoiding hitting rocks or being overturned going down a rapid.

The characteristics of the three views for different domains are shown in Table 9.1. The first activity is cycling, which despite being an everyday phenomenon has all the characteristics of being in control of a dynamic situation – as any novice cyclist can testify. In the case of cycling there is pretty little support to be had; the main reliance is on human perception and remembering, i.e., on the natural abilities. The only support is actually found in the view of the future, where traffic signs and signals may provide some help. (Obviously, seeing the traffic around and hearing specific sounds is also helpful, both for the present and the future views.) Driving a car is quite similar in terms of the nature of the process, but the stakes are greater. The speed is higher and there are more risks (less margin for failure). There are also a rapidly growing number of supports for the view of the present and the future, but not for the past. Here drivers still have to rely on what they – or their passengers – can remember.

Table 9.1: The Three Views for Different Processes

	View of the past	View of the present	View of the future
Cycling	Individual human memory	Unaided vision & hearing	Unaided vision, traffic signs and signals
Car driving	Human memory (individual, shared)	Aided vision, windscreen, instrumentation	Vision, traffic signs, communication, instrumentation
Aviation	Logs, memory (individual, shared)	Instrumentation, communication	ATC, FMS, instrumentation
Power generating stations	Charts, logs, trend diagrams, shift hand-over	Control room instrumentation	Computerised support, procedures

For two of the processes mentioned in Table 9.1, aviation and power generation, the conditions are quite different. There is first of all considerable support for the view of the past, often as an institutionalised part of the working environment relying on instrumentation and computing technology. This goes for the view of the present and the future as well. In flying, instrumentation and radio communication plays a significant role for both the present and the future. The latter is particularly important, since pilots presently fly according to directions from air traffic control rather than as they please. (This is, however, likely to change in the near future, but that is another story.) In process control proper, the view of the past is often institutionalised by necessity, because the process goes on over long periods of time, often months or years. Knowing what has happened in the past, on the previous shift or since start-up, can therefore be crucial. The view of the present is, of course, supported by control room design in itself, while the view of the future is given by procedures and instructions, and in some cases also by specific computerised support and displays.

Adaptation

While discussing the issues in the design of control rooms, even if some of them are virtual as in the above, it is nearly unavoidable to include the issue of adaptation, defined as the ability to make appropriate responses to changed or changing circumstances. Adaptation, and especially the concept of an adaptive interface or adaptive information presentation, has for many years been seen as the solution to many of the problems that have plagued process control. Indeed, the answer to the questions of presenting the right information, in the right form, at the right time, has often been taken to be adaptation. This would at least solve the problem of presenting the information in the right form. The argument for doing this is, however, slightly misleading. The mistake is that the ability *post hoc* to reason what the situation ought to have looked like is different from the ability to determine it as and when it happens. The belief in adaptation as a panacea nevertheless dies hard, and is still being promoted as the solution to the ever increasing volume of data in which we are immersed. A recent example is that the solution to the problem of too many information sources in a car, where drivers notably have to focus primarily on the driving, is to develop an adaptive interface. In human-machine interaction adaptation is typically proposed as one of the following.

- Changes to information presentation, which can be applied to contents, format, or timing. If this works, the obvious advantage is that it may facilitate signal discrimination, situation assessment and feedback evaluation. The problem and potential disadvantage are that it is very difficult to predict needs and conditions ahead of time, as discussed in Chapter 4.
- Response recognition and/or completion, which can be applied to single responses or to sequences of actions. This is certainly feasible, and a number of techniques are available, going back to Markov chains. The disadvantage is that it works best for command-type responses (single

responses), since the uncertainty of complex responses (sequences) can be formidable.

The view on and the attitude to adaptation reflects the assumptions of the underlying model of human performance. A procedural prototype model suggests that adaptation is beneficial because it reduces demands to 'input processing'. A contextual control model suggests that adaptation can be detrimental because it reduces predictability. Since predictability is of primary importance for JCSs, it stands to reason that adaptation should be used rather sparingly, if at all. The strength of humans is that they rather quickly can learn or find out how a system or process functions and use this to tailor their performance. If that process is degraded because the system keeps changing, the result may be oscillations or hunting due to insufficient damping. This is clearly not a desirable state.

Proponents of adaptive interfaces or adaptive interaction often point to the flexibility and efficiency of human-human communication. It is beyond dispute that humans are very capable of adapting the way they communicate to others, and that this – in most cases – is one of the reasons why communication is so effective. But the reason why this is possible is that humans – being models of each other – have the requisite variety built-in, so to speak. Communication is very much a case of being in or establishing control, as pointed out by Shannon & Weaver (1969; org. 1949). And control, as we know by now, requires a good model of the system to be controlled. Since artefacts are not good models of users – and are unlikely to become that for the foreseeable future – it follows that adaptation is doomed to fail as a panacea. This does not preclude that it may work under very specific circumstances, for instance well-understood and narrowly defined work contexts. Attempts to use it outside of such circumstances are nevertheless unlikely to succeed and may, at best, unwittingly impose the forced automaton metaphor.

DECISION SUPPORT

Another major field of application for CSE is decision making and decision support. One of the first books that clearly bore the mark of CSE was, indeed, about intelligent decision support systems (Hollnagel, Mancini & Woods, 1986). Neither the interest for, nor the importance of, this topic have waned in the years since then. On the contrary, the veritable explosion in computing and communication technologies has made the problem of decision making all the more potent.

Decision making has traditionally been about selecting among alternatives, i.e., choosing what to do. This tradition can be traced back to the roots of decision theory and the philosophical aspects of knowing what is correct and what is incorrect. The main body of decision research has adopted a rather formal or normative view (Edwards, 1954) and striven to find ways of reconciling the obvious inability of humans to behave as rational decision makers with the requirements of the theories (Gilovic et al., 2002). Despite two small revolutions – the principle of approximate decisions from the theory of satisficing (Simon, 1955) and the school of naturalistic decision making (Klein et al., 1993) – decision making is still very much seen as a question of making the right decision, hence of obtaining and processing information. The view that decision making is a distinct process that can and should be supported is a misunderstanding that has been inherited from normative decision theory and reinforced by the school of human information processing. That view should be compared to the more descriptive or naturalistic approaches where decision making is seen as sense making, and where therefore one should support making sense rather than making decisions. This can also be formulated as the question of whether decision making is something that occurs at a separate point in time or whether it is part of a continuous control process. In the latter case it becomes more important to support, e.g., monitoring, detection and recovery rather than to support decision making as such.

In a CSE perspective, decision making is not so much about *what* to do but about *how* and *when* it should be one. Decision-making is thus typically concerned with when to do something, about magnitude or how much should be done (level of force, amount of resources spent), about modes and means of implementation, etc. In other words, decision-making more often deals with the ways to carry out an alternative than the choice of the alternative as such. This is particularly so for situations where the alternatives are obvious and given in advance, often as binary choices, but where there can be many ways of implementing a specific alternative. A simple example is fire fighting: here you can either decide to fight the fire or to let it burn. But the decision to fight the fire requires further decisions about how to do it, how many resources to put in, which strategy to apply, when to do what, etc.

As a consequence of this, the nature of decision support changes and must be considered anew. First of all it cannot simply be an issue of automation, since decisions cannot be automated without ceasing being decisions. Automation works fine for routine tasks, where most – if not all – conditions can be anticipated. Automation requires that the environment is highly regular, i.e., that there is only a limited set of possible conditions, and that these can be identified with high reliability. But if the environment is highly regular then it is possible to plan in advance what to do. There is therefore no need to make decisions and consequently no need of a decision support system. Conversely, if the environment is irregular and unpredictable, then it is impossible to introduce effective automation. Decisions are required when the conditions are irregular, and when it is not known in advance how the system should respond. Decision support should therefore not be seen as automation in the usual meaning of the term, because it is not feasible to automate something that is not highly regular and reliable. Decision support does not automate what humans do, in the sense of taking over – as in prosthetic support. Decision support rather provides functions that humans cannot accomplish – or at least cannot accomplish well – and by virtue of that amplifies or augments human decision-making as an overall function.

Decision support must also be continuous rather than discrete, and closely integrated with the task. It means that decision support no longer can be treated as a separate issue, just as the decision no longer is a separate and identifiable process. In that sense many or all aspects of interface and interaction design become issues of decision support, but not as decision support in a separate capacity. In considering the difference between intelligent decisions or intelligent support, CSE makes clear that the intelligence clearly cannot be in the support but must be in the decision maker, i.e., in the controlling system. We should therefore strive to support intelligent decisions and the intelligent implementation of choices made. The implementation can be described as an issue of remaining in control of the situation, which essentially means avoiding unpredictable developments and events. In that sense the implementation issues (e.g., *how* and *when* rather than *what*) become issues of maintaining control in the face of difficulties. The design of decision support may therefore be approached from an analysis of the control issues, specifically how control can be lost and how it can be maintained and regained.

THE LAST WORDS

This book started by characterising the driving forces that have shaped CSE. These were computerisation and the growing complexity of systems that resulted from that, the increased conspicuousness of the human factor in the functioning – and malfunctioning – of these systems, and finally the effects of the dominant scientific paradigm, which promoted the view of humans as information processing systems. Of the three driving forces the first remains with us and is probably beyond control, even for the most self-assured Luddite. The second is less unalterable, since it very much depends on the perspective taken. That, in turn, is partly a consequence of the third driving force, which definitely is something that can be changed.

It is undeniable that thinking of humans – and of systems in general – in terms of information processing and transmission has been of immense value as a way of coping with the complexity of the real world. This is so both for the behavioural sciences, including human factors, for economics, for sociology, for genetics, and probably for many other sciences as well. Yet for the behavioural sciences the analogy has been partly devastating as well, both because it imposed a reduced description or model of humans and because it introduced a conceptual separation between humans and machines. As far as the former goes, this resulted in an unfortunate separation of cognition, or 'cold cognition' (Abelson, 1963) from other facets of human behaviour, particularly those that have to do with affect and emotion ('hot cognition'). One consequence of that is a misguided attempt to re-establish the balance by focusing on the emotional aspects, as if emotion and cognition were opposites. As far as the latter goes, this resulted in an uncritical acceptance of linear decomposition as the primary analytical principle. Although decomposition is necessary for analysis, it is crucial to keep in mind that the components found in this way are artefacts of the underlying system description, and that they are meaningful only in relation to the whole form which the analysis started.

The view espoused by CSE is based on the general systems perspective, which – as a paradigm – precedes information processing. This view is cogently represented by cybernetics, but the roots go back to Ludwig von Bertalanffy's General Systems Theory, and even further back to the 19th-century English philosopher of science George Henry Lewes, who suggested the important distinction between resultant and emergent phenomena.

CSE can in a shorthand version be said to deal with how joint cognitive systems use artefacts to cope with complexity. The emphasis of this formulation is on the dynamic aspects, on *what* the JCS does rather than on *how* it does it. The descriptions that have been developed by CSE, the concepts and the methods, have a pragmatic purpose, and their possible success must be seen by how well they achieve that purpose, i.e., how well they enable us to engineer cognitive systems. Although this book mostly has referred to human-machine systems, to operators working with technology, CSE is about joint cognitive systems in general, and not about humans only. This is another reason why there is little emphasis on the possible 'mechanisms of the mind' since the 'mind' of an organisation naturally is different from the 'mind' of an individual.

The first three chapters of this book have described the background of CSE as well as the basic conceptual and methodological constituents. The following three chapters went into the details of the three main themes – coping with complexity, use of artefacts, and joint cognitive systems. Chapters seven and eight presented the basic principles for modelling and understanding control, and for how the temporal dynamics could be described. The final chapter outlined how CSE can be applied, but for reasons of space this was necessarily kept short. A more extensive description of the application of CSE must go into details of how to observe and study JCSs in the field, how to derive the basic findings or control 'laws' that determine the behaviour and performance of JCSs; and last but not least how to design JCSs that work. That will, indeed, be the focus of a second book by the same authors, which tentatively is entitled *Joint Cognitive systems: Patterns in Cognitive Systems Engineering*.

CSE is not the first attempt to describe how humans cope with the complexity and unpredictability of their environment, nor will it be the last. We do believe that it offers a useful way of approaching these issues, and that it has the virtue of striving to be as simple as possible, in agreement with the tenets of the *Law of Parsimony*. For CSE this has the double meaning that it applies to the solutions that cognitive systems use to cope with complexity, as well as to our descriptions of how this is done. It is therefore appropriate that the last words are the Latin rendering of William of Occam's principle:

Pluralitas non est ponenda sine neccesitate

Bibliography

Abelson, R. P. (1963). Computer simulation of "hot" cognition. In S. S. Tomkins & S. Messick (Eds.), *Computer simulation of personality*. New York: John Wiley & Sons.

Allen J. (1983). Maintaining knowledge about temporal intervals. *Communications of the ACM, 26*, 832-843.

Alm, H. & Nilsson, L. (2001). The use of car phones and changes in driver behaviour. *International Journal of Vehicle Design, 26*, 4-11.

Alty, J. L., Elzer, P., Holst, O., Johannsen, G. & Savory, S. (1985). *Literature and user survey of issues related to man-machine interfaces for supervision and control systems*. Copenhagen, Denmark: Computer Resources International (ESPRIT P600 Final Report).

Alty, J. L., Khalil, C. J. & Vianno, G. (2001). Adaptable Interfaces in Process Control: Experience gained in the AMEBICA project, In M. Lind, (Ed.) *Proc. of the XX Annual Conf. On Human Decision Making and Manual Control*, 47-62.

Amalberti, R. & Deblon, F. (1992). Cognitive modeling of fighter aircraft's process control: A step towards an intelligent onboard assistance system. *International Journal of Man-Machine Studies, 36*, 639-671.

Annett, J. (2003). Hierarchical task analysis. In E. Hollnagel (Ed.), *Handbook of cognitive task analysis*. Mahwah, NJ: Lawrence Erlbaum Associates.

Arbib, M. A. (1964). *Brains, machines and mathematics*. New York: McGraw-Hill.

Aristotle (350 B.C.) *Nicomachean ethics* (translated by W. D. Ross). http://classics.mit.edu/Aristotle/nicomachaen.html (accessed 2004-09-02).

Ashby, W. R. (1956). *An introduction to cybernetics*. London: Methuen & Co.

Attneave, F. (1959). *Applications of information theory to psychology: A summary of basic concepts, methods, and results*. New York: Holt, Rinehart & Winston.

Bainbridge, L. (1983). Ironies of automation. *Automatica, 19*(6), 775-779.

Batchelder, S., Rizzo, M. Vanderleest, R. & Vecera, S. P. (2003). Traffic scene related change blindness in older drivers. In: *Proceedings of 2nd International Driving Symposium on Human Factors in Driver Assessment, Training and Vehicle Design*, July 21-24, 2003, Park City, Utah.

Beer, S. (1964). *Cybernetics and management*. New York: Science Editions.

Beer, S. (1981). *The brain of the firm* (2nd Ed). New York: Wiley.

Beer, S. (1985). *Diagnosing the system for organizations*. Chichester: Wiley.

Bergan, M. & Alty, J. L. (1992). Multimedia interface design in process control. *IEE Colloquium on Interactive Multimedia: A Review and Update for Potential Users*, 9/1-9/6.

Billings, C. E. (1996). *Aviation automation. The search for a human centered approach*. Hillsdale, N.J: Erlbaum.

Blom, H. A. P., Stroeve, S., Daams, J. & Nijhuis, H. B. (2001). Human cognition performance model based evaluation of air traffic safety. In S. W. A. Dekker, (Ed.), *Proceedings of the 4ᵗʰ International Workshop on Human Error, Safety and Systems Development*, Linköping, Sweden, 11-12 June 2001.

Bødker, S. (1996). Applying activity theory to video analysis: How to make sense of video data in HCI. In B. A. Nardi (Ed.), *Context and consciousness: Activity theory and human-computer interaction*. Cambridge, MA: MIT Press. (Pages 147-174.)

Boring, E. G. (1946). Mind and mechanism. *The American Journal of Psychology*, *59*(2), 173-192.

Broadbent, D. (1958). *Perception and communication*. London: Pergamon Press. C7,

Broadbent, D. E. (1977). Levels, hierarchies, and the locus of control. *Quarterly Journal of Experimental Psychology*, *29*, 181-201.

Broadbent, D. E. (1980). The minimization of models. In A. J. Chapman & D. M. Jones (Eds.), *Models of man*. Leicester: The British Psychological Society.

Broughton, J. & Baughan, C. (2002). The effectiveness of antilock braking systems in reducing accidents in Great Britain. *Accident Analysis & Prevention*, *34*, 347-355.

Bruner, J. S., Goodnow, J. J. & Austin, G. A. (1956). *A study of thinking*. New York: John Wiley & Sons, Ltd.

Brunswik, E. (1956). *Perception and the representative design of psychological experiments*. Berkeley: University of California Press.

Bye, A. Hollnagel, E. & Brendeford, T. S. (1999). Human-machine function allocation: A functional modelling approach. *Reliability Engineering and Systems Safety, 64,* 291-300.

Cacciabue, P. C. & Hollnagel, E. (1992). Simulation of cognition: Applications. In J.-M. Hoc, P. C. Cacciabue & E. Hollnagel (Eds.), *Expertise and technology: Cognition and human-computer cooperation*. New York: Lawrence Erlbaum Associates.

Card, S., Moran, T. & Newell, A. (1983). *The psychology of human-computer interaction*. Hillsdale, NJ: Erlbaum.

Carroll, J. M. & Campbell, R. L. (1988). *Artifacts as psychological theories: The case of human-computer interaction.* User Interface Institute, IBM T. J. Watson Research Centre, Yorktown Heights, New York.

Carver, C. S. & Scheier, M. F. (1982). Control theory: A useful conceptual framework for personality-social, clinical, and health psychology. *Psychological Bulletin*, *92*(1), 111-135.

Chapanis, A. (1970). Human factors in systems engineering. In K. B. De Greene (Ed.), *Systems psychology*, p. 51-78. New York: McGraw-Hill.

Cherry, E. C. (1953). Some experiments on the recognition of speech, with one and with two ears. *Journal of the Acoustic Society of America*, *25*, 975-979.

Cherry, E. C. (1957). *On human communication*. Cambridge, Massachusetts: The M. I. T. Press.

Clancey, W. J. (1992). Representations of knowing: In defense of cognitive apprenticeship. *Journal of Artificial Intelligence in Education*, *3*(2), 139-168.

Clark, A. 1997. *Being there*. Cambridge, MA: MIT Press.

Colburn, T. R. (1991). Program verification, defeasible reasoning, and two views of computer science. *Minds and Machines*, *1*, 97-116.

Collins, A. & Quillian, M. R. (1969). Retrieval time from semantic memory. *Journal of Verbal Learning & Verbal Behavior*, *8*, 240-247.

Colquhoun, R. (1984). Development of symptom-oriented operating procedures. *Nuclear Safety*, 25(3), May-June.

Conant, R. C. & Ashby, W. R. (1970). Every good regulator of a system must be a model of that system. *International Journal of Systems Science*, *1*(2), 89-97.

Cook, R. I. & Woods, D. D. (1996). Adapting to new technology in the operating room. *Human Factors*, *38*(4), 593-613.

Corcoran, W. R., Finnicum, D. J., Hubbard III, F. R., Musick, C. R. & Walzer, P. F. (1981). Nuclear Power Plant Safety Functions. *Nuclear Safety*, *22*(2).

Corcoran, W. R., Porter, N. J., Church, J. F., Cross, M. T. & Guinn, W. M. (1981). Critical safety functions. *Nuclear Technology*, *55*, 690-712.

Corker, K. & Verma, S. (2000). Introduction of context in human performance model as applied to dynamic resectorization. In *Proceedings of the 11th International Symposium on Aviation Psychology*, Columbus, Ohio, March 5-7, 2000.

Craik, K. J. W. (1943). *The nature of explanation*. Cambridge: Cambridge University Press.

Craik, K. J. W. (1947). Theory of the human operator in control systems - I. The operator as an engineering system. *British Journal of Psychology*, *38*, 56-61.

Craik, K. J. W. (1948). Theory of the human operator in control systems - II. Man as an element in a control system. *British Journal of Psychology*, *38*, 142-148.

Crovitz, H. F. (1970). *Galton's walk: Methods for the analysis of thinking, intelligence, and creativity*. New York: Harper & Row.

De Mey, M. (1982). *The cognitive paradigm*. Dordrecht: Reidel.

Decortis, F. & De Keyser, V. (1988). *Time: The Cinderella of man-machine interaction*. IFAC/IFIP/IEA/IFORS Conference on Man-Machine Systems, 14-16 June 1988, Oulu, Finland.

Dekker, S. W. A. & Hollnagel, E. (Eds.). (1999). *Coping with computers in the cockpit*. London: Ashgate.

Dekker S. W. A. & Hollnagel, E. (2004). Human factors and folk models. *Cognition, Technology & Work*, 6(2), 79-86/

Dekker, S. W. A. & Woods, D. D. (1999). To intervene or not to intervene: The dilemma of management by exception. *Cognition, Technology & Work*, 1(2), 86-96.

Dewey, J. (1896). The reflex arc concept in psychology. *Psychological Review*, 3(4), 357-370.

Donders, F. C. (1969). On the speed of mental processes. *Acta Psychologica*, 30, 412-431. (Translated from Over de snelheid van psychische processen. Onderzoekingen gedaan in het Physiologisch Laboratorium der Utrechtsche Hoogeschool, 1868-1869, Tweede reeks, II, 92-120.)

Dörner, D. (1980). On the difficulties people have in dealing with complexity. *Simulation & Games*, 11(1), 87-106.

Dreyfus, S. E. & Dreyfus, H. L. (1980). *A five-stage model of the mental activities involved in directed skill acquisition*. Operations Research Center, ORC-80-2. Berkeley, CA: University of California.

Duncker, K. (1945). On problem solving. *Psychological Monographs*, 58(270).

Edwards, W. (1954). The theory of decision making. *Psychological Bulletin*, 51(4), 380-417.

Endsley, M. R. (1995). Toward a theory of situation awareness in dynamic systems. *Human Factors*, 37, 32-64.

Fikes, R. E. & Nilsson, N. (1971). STRIPS: A new approach to the application of theorem proving to problem solving. *Artificial Intelligence*, 5(2), 189-208.

Fitts, P. M. (Ed). (1951). *Human engineering for an effective air navigation and traffic-control system*. Columbus, OH: Ohio State University Research Foundation.

Furukawa, H. & Inagaki, T. (1999). Situation-adaptive interface based on abstraction hierarchies with an updating mechanism for maintaining situation awareness of plant operators. *IEEE SMC '99 Conference Proceedings Systems, Man, and Cybernetics, Vol. 3*, 693-698.

Furuta, K., Sasou, K., Kubota, R., Ujita, H., Shuto, Y. & Yagi, E. (2000). Human factors analysis of JCO criticality accident. *Cognition, Technology & Work*, 2(4), 182-203.

Gallanti, M., Gilardoni, L., Guida, G., Stefanini, A. & Tomada, L. (1988). Integrating shallow and deep knowledge in the design of an on-line process monitoring system. In E. Hollnagel, G. Mancini & D. D. Woods (Eds.), *Cognitive engineering in complex dynamic worlds*. London: Academic Press.

Gibson, J. J. (1979). *The ecological approach to visual perception*. Boston: Houghton Mifflin.

Gibson, J. J. & Crook, L. E. (1938). A theoretical field-analysis of automobile-driving. *The American Journal of Psychology, LI*(3), 453-471.

Goldstine, H. (1972). *The computer from Pascal to von Neumann*. Princeton, New Jersey: Princeton University Press.

Goodstein, L. P. (1981). *Overview on control room design*. Roskilde, Denmark: Electronics Department (NKA/KRU-P2 (81) 42).

Grote, G., Weik, S., Wäfler, T. & Zölch, M. (1995). Complementary allocation of functions in automated work systems. In Y. Anzai, K. Ogawa & H. Mori (Eds.), *Symbiosis of human and artifact*. Amsterdam: Elsevier.

Hall, A. D. & Fagen, R. E. (1968). Definition of system. In W. Buckley (Ed.), *Modern systems research for the behavioural scientist*. Chicago: Aldine Publishing Company.

Hammond, R. R. (1993). Naturalistic decision making from a Brunswikian viewpoint: Its past, present, future. In G. A. Klein, J. Orasanu, R. Calderwood & C. E. Zsambok (Eds.), *Decision making in action: Models and methods*. Norwood, NJ: Ablex.

Haugeland, J. (1985). *Artificial intelligence: The very idea*. Cambridge, MA: MIT Press.

Helander, M. G., Landauer, T. K., Prabhu, P. & Prabhu, P. C. (1997). (Eds.), *Handbook of human-computer interaction* (2nd Ed.). New York, Elsevier Science Inc.

Hiltz, S. R. & Turoff, M. (1985). Structuring computer-mediated communication systems to avoid information overload. *Communications of the ACM, 28*(7), 680-689.

Hoc, J.-M., Amalberti, R. & Boreham, N. (1995). Human operator expertise in diagnosis, decision-making, and time management. In J.-M. Hoc, P. C. Cacciabue & E. Hollnagel (Eds.), *Expertise and technology*. Hillsdale, NJ: Lawrence Erlbaum.

Hollnagel, E. (1983). What we do not know about man-machine systems. *International Journal of Man-Machine Studies, 18,* 135-143.

Hollnagel, E. (1992). Coping, coupling and control: The modelling of muddling through. In P. A. Booth & A. Sasse (Eds.), *Mental models and everyday activities*. Proceedings of Second Interdisciplinary Workshop on Mental Models, March 23-25, Cambridge, UK.

Hollnagel, E. (1993a). *Human reliability analysis. Context and control.* London: Academic Press.

Hollnagel, E. (1993b). Models of cognition: Procedural prototypes and contextual control. *Le Travail humain, 56(1),* 27-51.

Hollnagel, E. (1993c). The phenotype of erroneous actions. *International Journal of Man-Machine Studies, 39,* 1-32.

Hollnagel, E. (1998a). *Cognitive reliability and error analysis method.* Oxford, UK: Elsevier Science.

Hollnagel, E. (1998b). Context, cognition, and control. In Y. Waern (Ed.). *Co-operation in process management - Cognition and information technology.* London: Taylor & Francis.

Hollnagel, E. (1999a). Modelling the controller of a process. *Trans. Inst. MC., 21(4),* 163-170.

Hollnagel, E. (1999b). From function allocation to function congruence. In S. Dekker & E. Hollnagel (Eds.), *Coping with computers in the cockpit.* Aldershot, UK: Ashgate.

Hollnagel, E. (2000). Modelling the orderliness of human action. In: N. B. Sarter & R. Amalberti (Eds.), *Cognitive engineering in the aviation domain.* Mahwah, NJ: Lawrence Erlbaum Associates.

Hollnagel, E. (2001). Extended cognition and the future of ergonomics. *Theoretical issues in Ergonomics Science,* 2(3), 309-315.

Hollnagel, E. (2003). *Handbook of cognitive task design.* Mahwah, NJ: Lawrence Erlbaum Associates.

Hollnagel, E. (2004). *Barriers and accident prevention.* Aldershot, UK: Ashgate.

Hollnagel, E., Cacciabue, P. C. & Rouhet, J.-C. (1992). *The use of integrated system simulation for risk and reliability assessment.* Paper presented at the 7th International Symposium on Loss Prevention and Safety Promotion in the Process Industry, Taormina, Italy, 4th-8th May, 1992.

Hollnagel, E., Mancini, G. & Woods, D. D. (Eds.) *Intelligent decision support in process environments.* Berlin-Heidelberg: Springer Verlag, 1986.

Hollnagel, E. & Niwa, Y. (2001). *Principles of performance monitoring in coupled human-machine systems.* Proceedings of 8th IFAC/IFIP/IFORS/IEA Symposium on Analysis, Design, and Evaluation of Human-Machine Systems. 18-20 September, Kassel, Germany.

Hollnagel, E. & Woods, D. D. (1983). Cognitive systems engineering. New wine in new bottles. *International Journal of Man-Machine Studies, 18,* 583-600

Hutchins, E. (1995). *Cognition in the wild.* Cambridge, MA: MIT Press.

Ihde, D. (1979). *Technics and praxis.* Dordrecht, Holland: D. Reidel.

James, W. (1981). *The Principles of Psychology,* Cambridge, MA: Harvard University Press, 1981. Originally published in 1890.

Jenkins, J. J. (1974). Remember that old theory of memory? Well, forget it! *American Psychologist, 29*, 785-795.

Johannsen, G. (1990). Fahrzeugführung. In C. G. Hoyos & B. Zimolong (Eds.), *Ingenieurpsychologie*. Göttingen, FRG: Verlag für Psychologie.

Johnson-Laird, P. N. (1980). Mental models in cognitive science. *Cognitive Science, 4*, 71-115.

Kawano, R., Ohtsuka, T. & Masugi, T. (2000). Awarable complexity : A study on CRT picture design based on Plant Images by NPP *Operators. In S. Kondo & K. Furuta (Eds.), PSAM*-5, Probabilistic safety assessment and management, Osaka, Japan.

Kirwan, B. (1994). *A guide to practical human reliability assessment.* London: Taylor & Francis.

Klein, G. A., Oramasu, J., Calderwood, R. & Zsambok, C. E. (1993). *Decision making in action: Models and methods*. Norwood, NJ: Ablex.

Klein, G. A., Ross, K. G., Moon, B. M., Klein, D. E., Hoffman, R. R. & Hollnagel, E. (2003). Macrocognition. *IEEE Intelligent Systems, May-June*, 81-85.

Kragt, H. (1983). *Operator tasks and annunciator systems*. Eindhoven, Holland: Eindhoven University of Technology.

Kruglanski, A. W. & Ajzen, I. (1983). Bias and error in human judgement. *European Journal of Social Psychology, 13*, 1-44.

Kuhn, T. S. (1970). The structure of scientific revolutions. *International Encyclopedia of Unified Science, 2*(2), (whole issue, 2nd ed.).

Landauer, T. K. (1975). Memory without organization: Properties of a model with random storage and undirected retrieval. *Cognitive Psychology, 7*, 495-531.

Lashley, K. (1950). In search of the engram. *Symposia of the Society for Experimental Biology, 4*, 454-482.

Leveson, N. G. (1993). An investigation of the Therac-25 accidents. *IEEE Computer, 26*(7), 18-41.

Leveson, N. G. (1995). *Safeware – system safety and computers*. Reading, MA: Addison-Wesley.

Lewin, K. (1936). *Principles of topological psychology*. New York: McGraw-Hill.

Lewin, K. (1951). Constructs in field theory. In D. Cartwright (Ed.), *Field theory in social science*. New York: Harper & Row.

Lewin, K. (1958). Group decision and social change. In Maccoby, E., Newcomb, T. M. & Hartley, E. L. (Eds.), *Readings in social psychology*. New York: Holt, Rinehart & Winston.

Lind, M. (1991). *On the modelling of diagnostic tasks*. 3rd European Conference on Cognitive Science Approaches to Process Control, Cardiff, September 2-6, 1991.

Lind, M. (2003). Making sense of the abstraction hierarchy in the power plant domain. *Cognition, Technology & Work, 5*(2), 67-81.

Lind, M. & Larsen, M. N. (1995). Planning support and the intentionality of dynamic environments. In J. M. Hoc, P. C. Cacciabue & E. Hollnagel (Eds.), *Expertise and technology: Cognition and human-computer interaction.* Hillsdale, NJ: Lawrence Erlbaum Associates.

Lindblom, C. E. (1959). The science of "muddling through." *Public Administration Quarterly, 19,* 79-88.

Lindsay, P .H. & Norman, D. A. (1976). *Human information processing.* New York: Academic Press.

Mackay, D. M. (1968). Towards an information-flow model of human behaviour. In W. Buckley (Ed.), *Modern systems research for the behavioral scientist.* Chicago: Aldine Publishing Company.

Mandler, G. (1975). *Mind and emotion.* New York: Wiley.

Maruyama, M. (1963). The second cybernetics: Deviation-amplifying mutual processes. *American Scientist, 55,* 164-179.

Marr, D. (1977) Artificial intelligence – A person view. *Artificial Intelligence, 9,* 37-48.

McConkie, G. W. & Currie, C. B. (1996). Visual stability across saccades while viewing complex pictures. *Journal of Experimental Psychology: Human Perception & Performance, 22*(3), 563-581.

McCormick, E. J. & Tiffin, J. (1974). *Industrial psychology.* London: George Allen and Unwin Ltd.

McRuer, D. T., Allen, W., Weir, D. H. & Klein, R. H. (1977). New results in driver steering control. *Human Factors, 19*(4), 381-397.

Michon, J. A. (1985). A critical view of driver behavior models. What do we know, what should we do? In L. Evans & R. Schwing (Eds.), *Human behavior and traffic safety* (pp. 485-525). New York: Plenum press.

Mihram, G. A. (1972). The modeling process. *IEEE Transactions on Systems, Man, Cybernetics, SMC-2*(5), 621-629.

Miller, G. A., Galanter, E. & Pribram, K. H. (1960). *Plans and the structure of behavior.* New York: Holt, Rinehart & Winston.

Miller, J. G. (1960). Information input overload and psychopathology. *American Journal of Psychiatry, 116,* 695-704.

Moray, N. (1959). Attention in dichotic listening: Affective cues and the influence of instructions. *Quarterly Journal of Experimental Psychology, 11,* 56-60.

Moray, N. (1967). Where is capacity limited? A survey and a model. In A. F. Sanders (Ed.), *Attention and performance 1.* Amsterdam: North-Holland Publishing Company.

Moray, N. (1970). *Attention.* New York: Academic Press.

Moray, N. (1981). *Human information processing and supervisory control.* Massachusetts Institute of Technology, Industrial Liaison Program.

Moray, N., Hiskes, D., Lee, J. & Muir, B. (1995). Trust and human intervention in automatic systems. In: J.-M. Hoc, P. C. Cacciabue, & E. Hollnagel (Eds.): *Expertise and technology. Cognition & human-computer cooperation*. Lawrence Erlbaum Associates.

Moray, N. & Inagaki, T. (2000). Attention and complacency. *Theoretical issues in Ergonomics Science*, *1*(4), 354-365.

Morick, H. (1971). Cartesian privilege and the strictly mental. *Philosophy and Phenomenological Research*, *31*(4), 546-551.

Muscio, B. (1921). Is a fatigue test possible? A report to the Industrial Fatigue Research Board. *British Journal of Psychology*, *12*, 31-46.

Neisser, U. (1967). *Cognitive psychology*. New York, Appleton Century Crofts.

Neisser, U. (1976). *Cognition and reality*. San Francisco: W. H. Freeman.

Neisser, U. (1982). *Memory observed. Remembering in natural contexts*. San Francisco: Freeman.

Newell, A. (1980). Physical symbol systems. *Cognitive Science*, *4*, 135-183.

Newell, A. & Simon, H. A. (1961). *GPS – A program that simulates human problem-solving*. In Proceedings of a Conference on Learning Automata, Technische Hochschule, Karlsruhe, April 11-14.

Newell, A. & Simon, H. A. (1972). *Human problem solving*. Englewood Cliffs, NJ: Prentice-Hall.

Newell, A. F. (1993). HCI for everyone. In S. Ashlund, A. Henderson, E. Hollnagel, K. Mullet & T. White (Eds.), *Proceedings of the INTERCHI '93 conference on Human factors in computing systems*. Amsterdam, The Netherlands: IOS Press.

Nielsen, J. (1993). *Usability engineering*. London: Academic Press.

Niwa, Y. & Hollnagel, E. (2001). Enhancing operator control by adaptive alarm presentation. *International Journal of Cognitive Ergonomics*, *5*(3), 367-384.

Norman, D. A. (1976). *Memory and attention* (2nd ed.). New York: Wiley.

Norman, D. A. (1981). Categorization of action slips. *Psychological Review*, *88*, 1-15.

Norman, D. A. (1983). *Position paper on human error*. The 2nd Clambake Conference on Human Error, Bellagio, Italy.

Norman, D. A. (1993). Things that make us smart: Defending human attributes in the age of the machine. Reading, MA: Addison-Wesley.

Norman, D. A. (1998). *The invisible computer*. The MIT Press.

Norman, D. A. & Draper, S. W. (1986). *User centered system design*. Lawrence Erlbaum.

O'Hara, J. M., Brown, W. S., Lewis, P. M. & Persensky, J. J. (2002). *The effects of interface management tasks on crew performance and safety in complex, computer-based systems: Overview and main findings*

(NUREG/CR-6690, Vol. 1). Washington, DC: U.S. Nuclear Regulatory Commission.

O'Regan, J. K. & Noë, A. (2001). A sensorimotor account of vision and visual consciousness. *Behavioral and Brain Sciences, 24*(5), 939-973.

Onken, R. & Feraric, J. P. (1997). Adaptation to the driver as part of a driver monitoring and warning system. *Accident Analysis & Prevention, 29*(4), 507-513.

Parasuraman, R. & Mouloua, M. (Eds.) (1996). *Automation and human performance. Theory and applications.* Mahwah, NJ: Lawrence Erlbaum Associates.

Parasuraman, R. & Riley, V. (1997). Humans and automation: Use, misuse, disuse, abuse. *Human Factors, 39*, 230-253.

Parasuraman, R. (1979). Memory loads and event rate control sensitivity decrements in sustained attention. *Science, 205*, 924-927.

Peddie, H., Filz, A. Y., Arnott, J. L. & Newell, A. F. (1990). *Extra-ordinary computer human operation (ECHO).* Presented at the 2nd Joint GAF/RAF/USAF Workshop on Electronic Crew Teamwork, Ingolstadt, Germany, 25-28 September, 1990.

Peirce, C. S. (1868). Some consequences of four incapacities. *Journal of Speculative Philosophy, 2*, 140-157. (Reprinted in P. P. Wiener (Ed.), *Charles S. Peirce - Selected Writings.* New York: Dover, 1958.)

Peterson, C. R. & Beach, L. R. (1967). Man as an intuitive statistician. *Psychological Bulletin, 68*(1), 29-46.

Pepper, S. C. (1942). *World hypotheses: A study in evidence.* University of California Press.

Perrow, C. (1984). *Normal accidents: Living with high-risk technologies.* New York: Basic Books.

Petroski, H. (1994). *Design paradigms: Case histories of error and judgment in engineering.* Cambridge: University of Cambridge Press.

Pew, R. W. & Barron, S. (1982). Perspectives on human performance modelling. *Proceedings of the IFAC Conference on Analysis, Design, and Evaluation of Man-Machine Systems,* Baden-Baden, FRG.

Petroski, H. (1994). Design paradigms. Case histories of error and judgment in engineering. Cambridge, UK: Cambridge University Press.

Pfaff, G. E. & ten Hagen, P. J. W. (Eds.), (1985). *User interface management systems.* Berlin: Springer Verlag.

Popper, K. R. (1959). *The logic of scientific discovery.* London: Hutchinson & Co Ltd.

Pringle, J. W. S. (1951). On the parallel between learning and evolution. *Behaviour, 3*, 174-215.

Rasmussen, J. (1981). *Human factors in high risk technology* (RISØ N-2-81). Roskilde, Denmark: Risø National Laboratories.

Rasmussen, J. (1986). *Information processing and human-machine interaction: An approach to cognitive engineering.* New York: North-Holland.

Rasmussen, J. & Jensen, A. (1974). Mental procedures in real-life tasks. A case study in electronic troubleshooting. *Ergonomics, 17,* 193-207.

Rasmussen, J. & Lind, M. (1981). *Coping with complexity* (Risø-M-2293). Roskilde, Denmark: Risø National Laboratory.

Rasmussen, J. & Vicente, K. J. (1987). *Cognitive control of human activities and errors: Implications for ecological interface design* (Risø-M-2660). Roskilde, Denmark: Risø National Laboratory.

Reason, J. T. (1979). Actions not as planned. The price of automatization. In G. Underwood & R. Stevens (Eds.), *Aspects of Consciousness,* Vol. 1. Psychological Issues. London: Wiley.

Reason, J. T. (1984). Absent-mindedness. In J. Nicholson & H. Belloff, (Eds.), *Psychology Survey No. 5.* Leicester: British Psychological Society.

Reason, J. T. (1986). *The classification of human error.* Unpublished manuscript. University of Manchester.

Reason, J. T. (1987). Generic error-modelling system (GEMS): A cognitive framework for locating human error forms. In J. Rasmussen, K. Duncan & J. Leplat (Eds.), *New technology and human error.* London: Wiley.

Reason, J. T. (1988). Cognitive aids in process environments: prostheses or tools? In E. Hollnagel, G. Mancini & D. D. Woods (Eds.), *Cognitive engineering in complex dynamic worlds.* London: Academic Press.

Reason, J. T. (1990). *Human error.* Cambridge, U.K.: Cambridge University Press.

Reason, J. T. (1997). *Managing the risks of organizational accidents.* Aldershot: Ashgate Publishing Limited.

Reason, J. T. & Mycielska, K. (1982). *Absent-minded? The psychology of mental lapses and everyday errors.* Englewood Cliffs, NJ: Prentice-Hall.

Reeves, B. & Nass, C. (2000). Perceptual bandwidth. *Communications of the ACM, 43*(1), 65-70.

Rich, E. (1983). Users are individuals: individualizing user models. *International Journal of Man-Machine Studies, 18,* 199-214.

Ritchie, B. F. (1953). The circumnavigation of cognition. *Psychological Review, 60,* 216-221.

Rochlin, G. I. (1986). High-reliability organizations and technical change: Some ethical problems and dilemmas. *IEEE Technology and Society Magazine, September,* 3-9.

Rosenblueth, A., Wiener, N. & Bigelow, J. (1968). Behavior, purpose and teleology. In W Buckley, (Ed.), *Modern systems research for the behavioral scientist.* Chicago: Aldine.

Roth, E. M., Woods, D. D. & Pople, H. E. Jr. (1992). Cognitive simulation as a tool for cognitive task analysis. *Ergonomics, 35,* 1163-1198.

Rouse, W. B. (1981). Human-computer interaction in the control of dynamic systems. *ACM Computing Survey, 13*(1), 71-99.

Rybczynski, W. (2000). *One good turn. A natural history of the screwdriver and screw.* Simon & Schuster UK Ltd.

Sagan, C. (1980). *Cosmos.* New York: Random House.

Sample, I. (2000). Bad vibrations. *New Scientist, No 2246*, p. 14.

Sanderson, P., Watanabe, J. & James, J. (1991). Visualization and analysis of complex sequential data using SHAPA (Mac). *Proceedings of the Third European Conference on Cognitive Science Approaches to Process Control.* Cardiff, UK. September 2-6, 1991.

Sarter, N. & Woods, D. D. (1997). Team play with a powerful and independent agent: Operational experiences and automation surprises on the Airbus A-320. *Human Factors, 39*(4), 553-569.

Schultz, D. P. (1975). *A history of modern psychology* (2nd ed.). New York. Academic Press.

Searle, J. R. (1980). Minds, brains, and programs. *The Behavioral and Brain Sciences, 3,* 417-424.

Seife, C. (1999). They have a problem. *New Scientist, No. 2218*, 14.

Senders, J. W. & Moray, N. (1991). *Human error. Cause, prediction, and reduction.* Hillsdale, NJ: Lawrence Erlbaum.

Shannon, C. E. & Weaver, W. (1969). *The mathematical theory of communication.* Urbana, Illinois: The University of Illinois Press.

Sheridan, T. B. (1992). *Telerobotics, automation, and human supervisory control.* Cambridge, MA: M. I. T Press.

Sheridan, T. B. (2002). *Humans and automation: System design and research issues.* New York, John Wiley & Sons, Inc.

Silverman, B. G. (1992). *Critiquing human error: A knowledge based human-computer collaboration approach.* London: Academic Press.

Simon, H. A. (1955). A behavioural model of rational choice. *The Quarterly Journal of Economics*, LXIX, 99-118.

Simon, H. A. (1967). Motivational and emotional controls of cognition. *Psychological Review, 74*(1), 29-39.

Simon, H. A. (1972). *The sciences of the artificial.* Cambridge, MA.: The M. I. T. Press.

Simon, H. A. (1979). *Models of thought.* Vol. 2. New Haven: Yale University Press.

Singleton, W. T. (1973). Theoretical approaches to human error. *Ergonomics, 16,* 727-737.

Smith, S. L. & Mosier, J. N. (1986). *Guidelines for designing user interface software.* Report ESD-TR-86-278. Bedford, MA: The MITRE Corporation.

Sobel, D. (1998). *Longitude.* London: Fourth Estate.

Sperling, G. A. (1963). A model for visual memory tasks. *Human Factors*, *5*, 1931.

Sperling, G. (1967). Successive approximations to a model for short term memory. In A. F. Sanders (Ed.), *Attention and performance I*. Amsterdam: North-Holland Publishing Company.

Stanton, N. A., Ashleigh, M. J., Roberts, A. D. & Xu, F. (2001). Testing Hollnagel's contextual control model: Assessing team behaviour in a human supervisory control task. *International Journal of Cognitive Ergonomics*, *5*(1), 21-33.

Stassen H. G., Johannsen G., Moray N., (1990), Internal representation, internal model, human performance model and mental workload, *Automatica*, *26*(4), 811-820.

Stassen, H. G. (1986). Decision demands and task requirements in work environments: What can be learned from human operator modelling. In E. Hollnagel, G. Mancini & D. D. Woods (Eds.), (1986). *Intelligent decision support in process environments*. Berlin: Springer Verlag.

Suchman, L. A. (1987). *Plans and situated actions. The problem of human-machine communication*. Cambridge, UK: Cambridge University Press.

Swain, A. D. (1982*). Modelling of response to nuclear power plant transients for probabilistic risk assessment*. Proceedings of the 8th Congress of the International Ergonomics Association. Tokyo, August, 1982.

Swain, A. D. (1990). Human reliability analysis: Need, status, trends and limitations. *Reliability Engineering and System Safety, 29*, 301-313.

Taylor, F. W. (1911). *The principles of Scientific Management*. New York: Harper & Row.

Taylor, J. R. (1993). *Risk analysis for process plant, pipelines and transport*. London, UK: Taylor & Francis.

Tufte, E. R. (2001). *The visual display of quantitative information*. Graphics Press.

Turing, A. M. (1936). On computable numbers, with an application to the *Entscheidungsproblem. Proceedings of the London Mathematics Society*, (ser. 2), *42*, p. 230-265.

Turing, A. M. (1950). Computing machinery and intelligence. *Mind*, October, *59*, 433-460.

Turk, M. & Robertson, G. (2000). Perceptual user interfaces. *Communications of the ACM, 43*(1), 33-34.

Tversky, A. & Kahneman, D. (1974). Judgment under uncertainty: Heuristics and biases. *Science, 185*, 1124-1131.

Vanderhaegen, F., Crévits, I., Debernard, S. & Millot, P. (1994). Human-machine cooperation: toward an activity regulation assistance for different air traffic control levels. *International Journal on Human-Computer Interaction, 6*, 65-104.

Vaughan, W. S. & Maver, A. S. (1972). Behavioural characteristics of men in the performance of some decision-making task component. *Ergonomics, 15*, 267-277.

Vicente, K. J. (1999). *Cognitive work analysis: Towards safe, productive, and healthy computer-based work.* Mahwah, NJ: Erlbaum.

Vicente, K. J., Christoffersen, K. & Pereklita, A. (1995). Supporting operator problem solving through ecological interface design. *IEEE Transactions on Systems, Man and Cybernetics, 25* (4), 529-545.

Vicente, K. J. & Rasmussen, J. (1992). Ecological interface design: Theoretical foundations. *IEEE Transactions on Systems, Man, Cybernetics, SMC-22*, 589-596.

Volta, G. Time and decision. In E. Hollnagel, G. Mancini & D. D. Woods (Eds.), (1986). *Intelligent decision support in process environments.* Berlin: Springer.

Waern, Y. & Cañas, J. J. (2003). Microworld task environments for conducting research on command and control. *Cognition, Technology & Work, 5*(3), 181-182.

Watson, J. B. (1913). Psychology as the behaviorist views it. *Psychological Review, 20*, 158-177.

Weir, G. & Alty, J. L. (Eds.) (1989). *Human-computer interaction and complex systems.* London: Academic Press.

Wäfler, T., Grote, G., Windischer, A. & Ryser, C. (2003). KOMPASS: A method for complementary system design. In E. Hollnagel (Ed.), *Handbook of cognitive task design.* Mahwah, NJ: Erlbaum.

Weick, K. E., Sutcliffe, K. M. & Obstfeld, D. (1999). Organizing for high reliability: Processes of collective mindfulness. *Research in Organizational Behavior, 21*, 13-81.

Weizenbaum, J. (1976). *Computer power and human reason.* San Francisco: W. H. Freeman and Company.

Westrum, R. (1993). Cultures with requisite imagination. In J. A. Wise, V. D. Hopkin & P. Stager (Eds.), *Verification ad validation of complex systems: Human factors issues.* Berlin: Springer Verlag. (pp. 401-416).

Wickens, C. D. (1984). *Engineering psychology and human performance.* Columbus, OH: Merrill.

Wickens, C. D. (1987). Information processing, decision-making, and cognition. In G. Salvendy (Ed.), *Handbook of human factors.* New York: John Wiley & Sons. (p. 72-107).

Wiener, E. L. (1988). Cockpit automation. In: E. L. Wiener & D. C. Nagel (Eds.), *Human factors in aviation.* San Diego, CA: Academic press.

Wiener, N. (1965). *Cybernetics.* Cambridge, Massachusetts: The M. I. T. Press.

Wiener, N. (1954). *The human use of human beings: Cybernetics and society.* Boston: Houghton Mifflin Co.

Wilde, G. J. S. (1982). The theory of risk homeostasis: Implications for safety and health. *Risk Analysis*, *2*(4), 209-225.

Winograd, T. (1972). *Understanding natural language*. New York: Academic Press.

Winograd, T. & Flores, F. (1986). *Understanding computers and cognition*. Reading, MA: Addison-Wesley.

Wood, G. (2002). *Living dolls. A magical history of the quest for mechanical life*. London: Faber and Faber.

Woods, D. D. (1986). Paradigms for intelligent decision support. In E. Hollnagel, G. Mancini & D. D. Woods (Eds.). *Intelligent decision support in process environments*. New York: Springer Verlag.

Woods, D. D. (1998). Designs are hypotheses about how artifacts shape cognition and collaboration. *Ergonomics*, *41*, 168-173.

Woods, D. D: (1995). Towards a theoretical base for representation design in the computer medium: Ecological perception and aiding human cognition. In J. Flach, P. Hancock, J. Caird & K. Vicente (Eds.), *An ecological approach to human machine systems I: A global perspective*. Erlbaum.

Woods, D. D. & Cook, R. I. (2002). Nine steps to move forward from error. *Cognition, Technology & Work*, *4*(2), 137-144.

Woods, D. D., Johannesen, L. J., Cook, R. I. & Sarter, N. B. (1994). *Behind human error: Cognitive systems, computers and hindsight*. WPAFB, OH: CSERIAC.

Woods, D. D. & Sarter, N. B. (2000). Learning from automation surprises and going sour accidents. In N. Sarter and R. Amalberti (Eds.). *Cognitive engineering in the aviation domain*. Hillsdale, NJ: Erlbaum.

Woods, D. D. & Watts, J. C. (1997). How not to have to navigate through too many displays. In Helander, M. G., Landauer, T. K. & Prabhu, P. (Eds.) *Handbook of human-computer interaction* (2nd edition), (pp. 617-650). Amsterdam: Elsevier Science.

Worm, A. (2001). Breaking the barriers: Facilitating efficient command and control in multi-service emergency management. *8th World Conference on Emergency Management* (TIEMS), June 19-22, Oslo, Norway.

Yoshida, H., Takeda, M., Hayashi, Y. & Hollnagel, E. (1997). Cognitive control behavior of the operator in the emergency. In P. Seppälä, T. Luopajärvi, C.-H. Nygård & M. Mattila (Eds.), *Proceedings of the 13th Triennial Congress of the International Ergonomics Association*, June 29-July 4, 1997, Tampere, Finland. Helsinki: Finnish Institute of Occupational Health.

Zapf, D., Brodbeck, F. C., Frese, M., Peters, H. & Prumper, J. (1990). Error working with office computers. In J. Ziegler (Ed) *Ergonomie und Informatik*. Mitteilungen des Fachausschusses 2.3 heft, 9, 3-25.

Author Index

Subject Index

Printed in the United States
by Baker & Taylor Publisher Services